Questions and Answers on

Thermoluminescence (TL) and Optically Stimulated Luminescence (OSL)

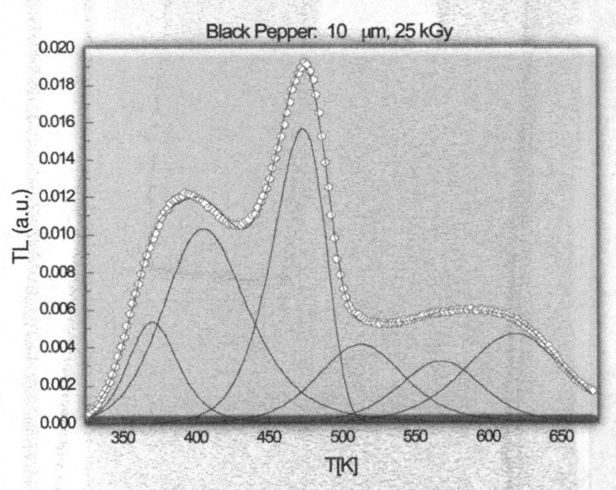

Questions and Answers on

Thermoluminescence (TL) and Optically Stimulated Luminescence (OSL)

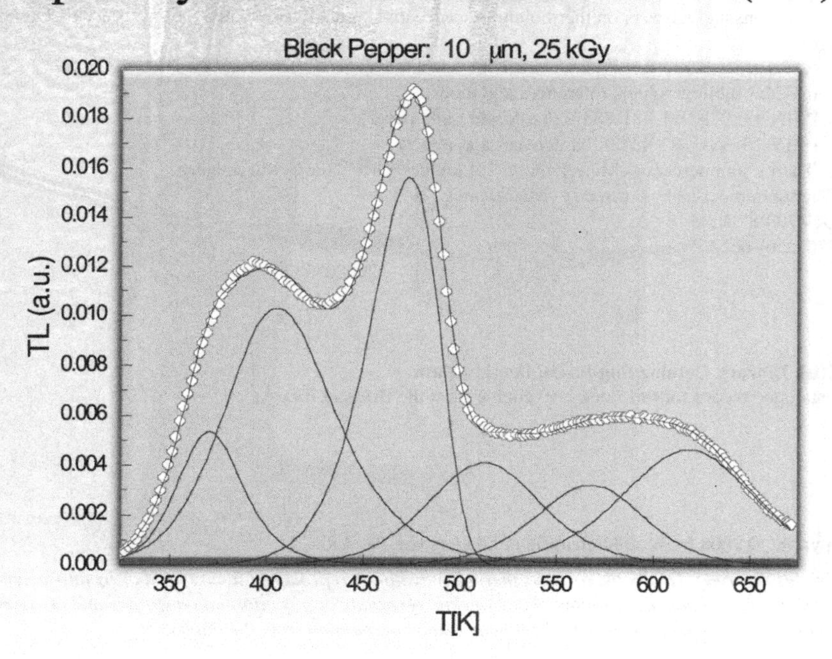

Claudio Furetta

Touro University Rome, Italy
Division of Touro College New York

NEW JERSEY • LONDON • SINGAPORE • BEIJING • SHANGHAI • HONG KONG • TAIPEI • CHENNAI

Published by

World Scientific Publishing Co. Pte. Ltd.

5 Toh Tuck Link, Singapore 596224

USA office: 27 Warren Street, Suite 401-402, Hackensack, NJ 07601

UK office: 57 Shelton Street, Covent Garden, London WC2H 9HE

Library of Congress Cataloging-in-Publication Data

Furetta, C., 1937-

Questions and answers on thermoluminescence and optically stimulated luminescence / Claudio Furetta.

p. cm.

Includes bibliographical references and index.

ISBN-13: 978-981-281-883-6 (hardcover : alk. paper)

ISBN-10: 981-281-883-9 (hardcover : alk. paper)

1. Thermoluminescence--Miscellanea. 2. Luminescence dating--Miscellanea. 3. Thermoluminescence dosimetry--Miscellanea. I. Title.

QC479.F87 2008

535'.356--dc22

2008033324

British Library Cataloguing-in-Publication Data

A catalogue record for this book is available from the British Library.

Typeset by Stallion Press

Email: enquiries@stallionpress.com

Printed in Singapore.

DEDICATION

I am deeply grateful to my wife Maria Clotilde for her constant and loving support to my scientific work. This book is dedicated to her.

PREFACE

This book on thermoluminescence and related phenomena, based on questions and answers, is born after many years of teaching.

The aim of the book is to cover not all but at least a great percentage of the possible questions which can arise during a teaching course as well as to solve quickly the doubts of the students during the study and the laboratory practice.

The book collects a series of questions, and the relative answers, over the main arguments of thermoluminescence, i.e. general properties, theory and kinetics methods. Other sections are concerning with thermoluminescence related phenomena as optically stimulated luminescence, luminescence dating, thermally stimulated conductivity, thermo stimulated current and so on. At the end of the book a large list of reference books and papers is also given.

The present book offers then a unique advantage to students and teachers in solid state physics, radiation dosimetry and nuclear science and techniques. The book, at the same time, is a strong support for the available thermoluminescence text books.

ACKNOWLEDGEMENTS

The author is grateful to Dr. Andrea Favalli, of the Ispra Research Centre, and to Dr. Eric Gauthier, for their sincere help.

CONTENTS

Chapter 1

GENERAL PROPERTIES OF THERMOLUMINESCENCE

Q 1. What is **radio-thermoluminescence (RTL)**?

A. It is the old name for thermoluminescence.

Q 2. How is it possible to decide if the **TL response vs dose** is linear or not?

A. Plot the TL emission against the dose on a log–log paper. We get linearity if the obtained straight line makes an angle of 45° with the logarithm axis.

Q 3. What is an **activator**?

A. It is a chemical element which is added to a thermoluminescent material for enhancing its luminescence efficiency.

Q 4. What is the **annealing** procedure?

A. The annealing procedure is a thermal treatment which erases from a thermoluminescent material any previous irradiation effects.

Q 5. What is the meaning of the term "**afterglow**"?

A. The term "afterglow" is used to indicate the luminescence emitted from a thermoluminescent material immediately after its irradiation. In general this phenomenon is temperature independent. A correlation between afterglow and a thermal fading has been observed.

Q 6. What is **anomalous fading**?

A. It is the phenomenon in which the trapped charges can escape from the traps at a rate much faster than the one expected from the calculated mean lifetime. It is weakly dependent on the temperature.

Q 7. What is the **calibration factor**?

A. The calibration factor, Φ_C, allows us to transform the light emission from a phosphor to the dose received by the phosphor itself. The calibration factor includes both reader and phosphor proprieties.

Q 8. How is the **calibration factor**, Φ_C, defined?

A. If we indicate with D_C a calibration dose and with M_C the light signal from a phosphor, corrected by the background, the calibration factor is given by the following expression

$$\Phi_C = \frac{D_C}{M_C}.$$

Q 9. How is the **phosphor sensitivity**, S, defined?

A. The sensitivity of a phosphor is generally expressed by the following expression:

$$S = \frac{light\ emission}{D \cdot m}$$

where $\boldsymbol{D}$ is the given dose, generally in the linear region of the calibration curve, and $\boldsymbol{m}$ is the mass of the phosphor.

Q 10. How do we set the **heating rate** for determining the kinetics parameters relative to a TL glow peak?

A. The heating rate has to set as slow as possible for the following reasons: (i) to obtain the best separation among the peaks (ii) to get the best thermal contact between the TL sample and the heating planchet.

Q 11. How can a **trap** be considered stable at room temperature?

A. A trap, having an activation energy E, is stable at room temperature if $\mathbf{E} \gg \mathbf{kT}$, where T is the room temperature value.

Q 12. What is the **optical bleaching** of the TL signal?

A. An irradiated material is exposed to visible light and, very often, the TL signal results to be reduced compared to the one obtained if the material had not been exposed to the light. This effect is due to eviction of charges from the trap owing to optical stimulation.

Q 13. What is the **light-induced fading**?

A. It is the other name of the optical bleaching of the TL signal (see Q12).

Q 14. What is the **phototransferred TL**?

A. This phenomenon indicates the transfer of charges from deep traps to shallower traps. It is caused by the light exposure of an irradiated sample.

Q 15. What is the **optically-induced TL**?

A. This effect concerns an unirradiated sample exposed to light: optical stimulation can allow transitions of electronic charge from pre-existing defects. The free charges so obtained can be trapped at empty trapping levels and give, then, a TL signal.

Q 16. What is the method called "**in vivo dosimetry**"?

A. It is a method to check the absorbed dose which has been delivered to a patient during radiological inspection or during therapeutic treatment.

Q 17. How do we perform the "**in vivo dosimetry**"?

A. It is performed by putting dosimeters on the patient's skin or in natural body cavities.

Q 18. What is the aim of "**in vivo dosimetry**"?

A. The aim of "in vivo dosimetry" is to check the target dose (ICRU 1993), in order to verify the correct delivery of irradiation to patients. In other words, the method allows us to compare the dose obtained from the detector's signal placed on the patient skin, with the theoretical value as calculated by the TPS.

Q 19. What is the meaning of **TPS**?

A. It refers to the Treatment Planning System. This system allows us to calculate the irradiation dose delivered to a patient during external radiotherapy.

Q 20. How are the **entrance dose**, $D_{entrance}$, and the **exit dose**, D_{exit}, defined?

A. The entrance and exit doses are defined at points a certain distance from the patient surface at the entrance and exit of the beam. This distance is equal to the depth of maximum build-up, d_{max}.

Q 21. How is the **surface dose**, $D_{surface}$, defined?

A. The surface dose is defined at 0.05 cm below the entrance surface.

Q 22. What is the intrinsic **precision** of a TL dosimeter?

A. The intrinsic precision is the reproducibility of a given thermoluminescent material associated with a given readout system.

Q 23. Which are the parameters that affect the **reproducibility** of a given thermoluminescent material associated to a given readout system?

A. The reproducibility is dependent on the quality of the thermoluminescent material, the annealing procedure, the reader characteristics, the thermal readout cycle and the purity of the nitrogen gas.

Q 24. How can the **reproducibility** evaluated?

A. It can be evaluated by randomly taking some thermoluminescent dosimeters of the same batch, i.e. 10 samples, and by irradiating them to a same dose. After readout and annealing treatment, the procedure is repeated several times. A standard deviation of $\pm 2\%$ or less is the index for a good reproducibility.

Q 25. What is the **thermal treatment** for thermoluminescent dosimeters?

A. The thermal treatment, usually called annealing, is a procedure performed in an oven before and after use of the dosimeters.

Q 26. Why is the **thermal treatment** important?

A. The thermal treatment or annealing allows us to stabilize the sensitivity and the background of the dosimeters in such a way that their dosimetric properties remain constant during the use.

Q 27. Which types of **heating systems** are encountered in TLD readers?

A. There are mainly two different systems: (i) heating by contact and (ii) non-contact systems.

Q 28. How does the **contact readout system** work?

A. The metallic support of the TL sample is heated by an electric current or by a hot finger moved by a lift mechanism.

Q 29. How does the **non-contact readout system** work?

A. The TL sample is heated by hot nitrogen gas, or by a laser beam, or by a light pulse from an halogen lamp.

Q 30. Which type of **heating cycle** has to be used for kinetics studies?

A. The most usual cycle is a linear one: the TL sample is progressively heated using the lowest possible heating rate.

Q 31. What is the definition of **fading**?

A. Fading is the spontaneous escape, at ambient temperature, of charge carriers from traps.

Q 32. What is the **fading factor**?

A. The fading factor is a parameter which allows us to calculate how much thermoluminescent information is lost per unit of time.

Q 33. How do we calculate the **fading factor**?

A. Considering the first order expression as a function of time, i.e.

$$\Phi = \Phi_0 \exp(-pt)$$

where

Φ is the peak area at the elapsed time t
Φ_0 is the peak at time $t = 0$
The fading factor p is given by

$$p = -\frac{1}{t} \ln\left(\frac{\Phi}{\Phi_0}\right).$$

Q 34. Is there an expression which takes into consideration the competition between **fading** and irradiation at the same time?

A. The expression is the following

$$\Phi = \frac{\dot{D}}{F_c s \exp\left(-\frac{E}{kT}\right)} \left\{1 - \exp\left[-ts \exp\left(-\frac{E}{kT}\right)\right]\right\}$$

where

$\dot{D}$ is the dose rate
F_c is the calibration factor of the TL system (dose/TL)
t is the observation time
T is the storage temperature.

Q 35. Is there a **desensitization** effect caused by UV exposure?

A. Yes. The thermoluminescent sensitivity can be increased artificially by irradiation with ionizing radiation or thermal treatment. The exposure to UV rays can reverse the sensitization effect of the previous procedure.

Q 36. Is there a thermoluminescence transferred effect as a consequence of the UV exposure?

A. Yes. The UV rays exposure can transfer some charges from deep traps to shallow ones, so that the peaks corresponding to the shallow traps, i.e. low temperature peaks, increase their sensitivity.

Q 37. What is the meaning of the term "**regeneration**" for a given peak?

A. It means an increase of the thermoluminescent intensity as a consequence of the charge transfer induced by UV exposure.

Q 38. Can a **pressure** applied on a thermoluminescent sample cause some effect?

A. The pressure application can excite valence electrons which can then be trapped or de-excite any trapped electron which can recombine. In both cases the thermoluminescence observed will be different from the one observed prior to the stress application.

Q 39. Which kind of **stress** applied to a sample can affect its thermoluminescent emission?

A. Stress can arise from operations such as crushing, grinding, packing and so on.

Q 40. Which are the possible **excitation agents** in thermoluminescence?

A. The excitation is achieved by conventional agents as ionizing radiations and, in some cases, also by UV rays.

Q 41. Is heat an **excitation agent** for thermoluminescence?

A. No. Heat is only a stimulating agent.

Q 42. How do we distinguish between **luminescence** and **incandescence** emissions?

A. Luminescence usually lies in a spectral region where the material of interest is not absorbing; incandescence occurs where absorptivity of the material is the maximum.

Q 43. Are there any differences in temperatures of occurrences between **thermoluminescence** and **incandescence**?

A. Yes. Incandescence occurs at very high temperatures, near the melting point of materials. The thermoluminescence occurs at much lower temperatures.

Q 44. Which are the most **sensitive thermoluminescent materials**?

A. The most sensitive thermoluminescent materials are the solid dielectrics. However, thermoluminescence is exhibited by minerals contained in inorganic crystals, glasses, ceramics, polymers and so on.

Q 45. How do we carry out **thermoluminescence spectra** measurements?

A.

- use band pass filters between sample and PMT while taking the glow curve
- periodical spectral scanning during glow curve measurements
- spectral scanning during isothermal decay measurements
- construction of the spectrum for any temperature of emission from the monochromatic thermoluminescent glow curve

Q 46. What is the effect of the **linear energy transferred** (**LET**) on the thermoluminescence response?

A. The effect of LET on the thermoluminescence response is, mostly, a decrease of thermoluminescence **sensitivity** with increase of LET of the incident radiation.

Q 47. What are the effects of high **LET** radiation on the thermoluminescence emission?

A. The main effect is a saturation of the thermoluminescent signal. Sometimes the LET effects manifest in changes of the glow curve, of the fading rate and on the thermoluminescent emission spectrum.

Q 48. Is there any intrinsic thermoluminescence **sensitivity** to UV rays exposure?

A. Upon UV excitation by an appropriate frequency, luminescence may be emitted during thermal stimulation, depending on the type of phosphor used.

Q 49. Are the **glow curves** obtained after UV rays exposure the same as the ones obtained after X or gamma irradiation?

A. The glow curves obtained after the two kinds of excitations are not usually the same.

Q 50. Is the **thermoluminescence sensitivity** to UV rays exposure the same as that of X or gamma irradiation?

A. In general the thermoluminescence sensitivity to UV rays exposure is very small compared to the one of X or gamma radiation.

Q 51. Is there a special procedure to enhance the **UV sensitivity** of phosphors?

A. Some materials, such as LiF, CaF_2, BeO and others, become UV sensitive or increase their intrinsic UV sensitivity after a high temperature treatment.

Q 52. Does the thermoluminescent glow curve shape depend on the **UV wavelength**?

A. Yes, the thermoluminescent glow curve shape is generally UV wavelength dependent.

Q 53. Is there a **UV effect** on phosphors already irradiated using X or gamma radiation?

A. If a sample has already been irradiated by X or gamma radiation, one may observe some bleaching effect in the ionizing radiation induced thermoluminescence. The bleaching effect is usually not the same for all the peaks, if there more than one in the glow curve, because each peak may have a different dependence for the bleaching effect.

Q 54. Which are the effects of the **heating rate** on the peak characteristics?

A. In the hypothesized situation of very good thermal contact between the thermoluminescent sample and the heating element, the following results are obtained:

- the peak temperature at the maximum of the peak, T_M, shifts to higher temperature values as the heating rate increases
- the thermoluminescent intensity, peak area and peak height decreases as the heating rate increases

The following figure shows the behavior of the two above characteristics as a function of the heating rate.

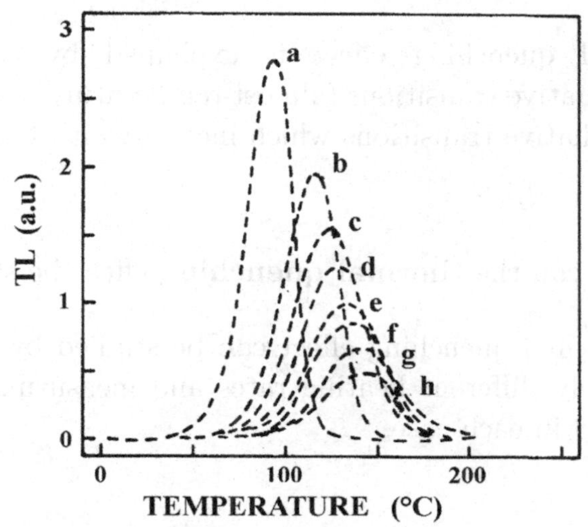

Change of the peak shape and shift in the peak position as a function of the heating rate. From (a) to (h) = 2, 8, 20, 30, 40, 50, 57, 71°C/s.

Q 55. Why does the thermoluminescent intensity decreases as the **heating rate** increases?

A. The thermoluminescent intensity decreases as the heating rate increases because of a thermal quenching effect, whose efficiency increases as the temperature increases, i.e. the glow peak shifts to higher temperature values as a consequence of the increase of the heating rate.

Q 56. What is the **thermal quenching** effect?

A. In general, the luminescence efficiency is a factor that is temperature dependent, so the efficiency decreases with increase the temperature.

Q 57. How can the **thermal quenching** effect be explained?

A. Thermal quenching effect is explained by competition between radiative transitions (almost temperature independent) and non-radiative transitions which increase with temperature.

Q 58. How can the **thermal quenching** effect be studied?

A. The thermal quenching effect can be studied by employing two extremely different heating rates and measuring the total light emitted in each case.

Q 59. Can an **electric field** have an effect on thermoluminescence when applied to a phosphor sample?

A. Some effects on the thermoluminescent emission have been observed during sample heating, i.e. to enhance the thermoluminescent emission.

Q 60. What is the order of magnitude of the **electric field** to be applied to a phosphor to get the effects?

A. The electric field should be of the order of 10^5 V/cm.

Q 61. What are the factors affecting the thermoluminescent emission when an **electric field** is applied to a phosphor?

A. It seems that the factors are:

- field ionization of the electron traps
- acceleration of electrons after their thermal release from traps and subsequent ionization

Q 62. What is the **detection threshold**?

A. The detection threshold is defined as the smallest dose that can be distinguished significantly from the zero dose.

Q 63. How do we calculate the **detection threshold**?

A. The detection threshold is calculated taking three times the standard deviation of the zero dose reading, expressed in units of the absorbed dose.

Q 64. In which way do we measure the **zero-dose**?

A. It is the signal obtained from a non-irradiated dosimeter, i.e. the dosimeter's background, L_{BKG}.

Q 65. How to calculate the **zero-dose** reading?

A. The zero-dose reading is the result of two components:

- the signal of the TL reader obtained without a dosimeter using the read out cycle used for a dosimeter reading. This quantity is called R_0 and it corresponds to the dark current of the TL system
- the signal from an annealed and unirradiated dosimeter, i.e. R_U

Several measurements of R_0 and R_U have to be performed. The zero-dose reading, or its average value, is then given by

$$\bar{R}_{BKG} = \bar{R}_0 + \bar{R}_U .$$

Q 66. How is the **net reading** of N dosimeters irradiated at the same dose defined?

A. It is defined as the difference between the mean value of the readings of the N irradiated dosimeters and the mean of their individual backgrounds. The following expression gives the net reading

$$R_{net} = \frac{1}{N}\left[\sum_{i=1}^{N} R_{irr,i} - \sum_{i=1}^{N} R_{BKG,i}\right] .$$

Q 67. How is it possible to repeat a **thermoluminescent experiment** reported in literature to get the same results?

A. The question has two different folders. The first one is concerned with the dosimetric characteristics of a given thermoluminescent material. Also, using the same kind of phosphor, it is normal to have differences in the thermoluminescent characteristics from batch to batch as well as within the same batch, from sample to sample. The only way to obtain similar results is to know exactly what are the experimental conditions used in literature. A second important factor is the TL instrument: in the case of using the same type of equipment, the electronic characteristics cannot be equal depending on the age, use and other factors affecting the electronic components. So, the dosimetric characteristics of a given material can be quite different from one laboratory to another. In principle, the only characteristics of a material which should be the same in any laboratory, everywhere, are the kinetics parameters, i.e. the activation energy, the frequency factor and the kinetics order, because they are independent of the instruments used.

Q 68. What is **fluorescence**?

A. Fluorescence is a luminescent process, a luminescent decay, which only persists for as long as the excitation is continued.

Q 69. Is **fluorescence decay** temperature dependent?

A. No, fluorescence decay is independent of temperature.

Q 70. What is the mechanism of **fluorescence**?

A. The fluorescence is determined by the transition probability from an excited level E_e to the ground state E_0. The following figure shows this effect.

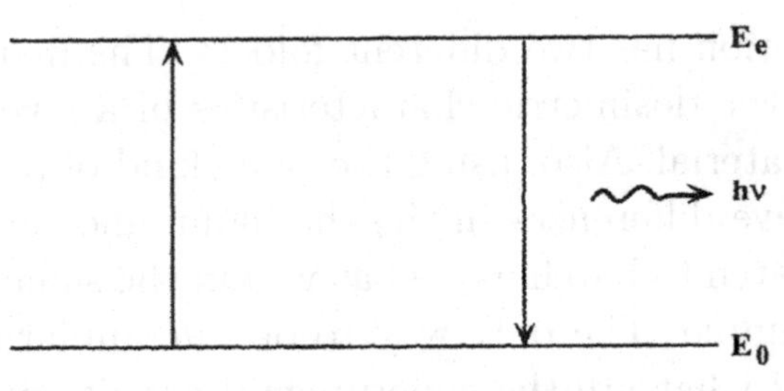

The fluorescent phenomenon.

Q 71. What is **phosphorescence**?

A. Phosphorescence is a luminescent phenomenon observable after removal of the exciting source.

Q 72. Is **phosphorescence** dependant on temperature?

A. Yes.

Q 73. Why does the **phosphorescence** depends on temperature?

A. This is because it can be delayed by some trapping levels E_m in the forbidden gap. An increase in temperature provides enough energy for detrapping. The following figure shows the phosphorescence phenomenon.

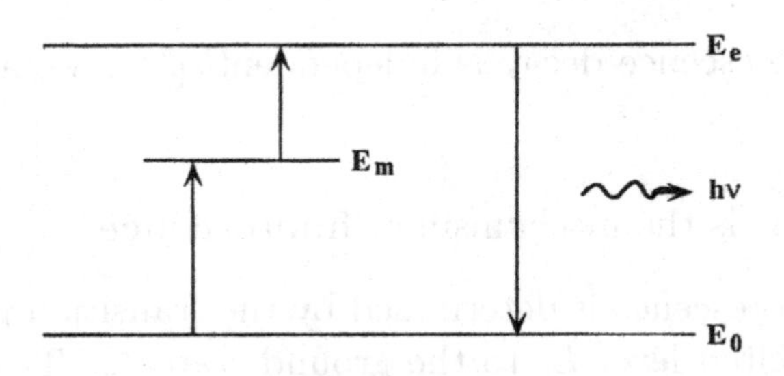

The phosphorescence phenomenon.

Q 74. How should the **spectrum** of the emitted thermoluminescent light be?

A. The spectrum of the emitted thermoluminescent light should be in the range where the detector system (photomultiplier and filter combination) responds at its best.

Q 75. Which glow peak characteristics are affected by the **heating rate**?

A. The heating rate, during a thermoluminescent measurement, affects the glow peak temperature position, the intensity of the peak and its shape.

Q 76. What is the main problem in the preparation of **artificial thermoluminescent materials**?

A. To find a very efficient activator impurity.

Q 77. Which are the main techniques for growing single **thermoluminescent crystals**?

A. The most important methods are those due to Bridgemann and Stockberger.

Q 78. In which way can **polycrystalline TL materials** be prepared?

A. An efficient method consists of precipitating the basic material and the activator, followed by drying and firing at high temperature in an inert atmosphere or in vacuum.

Q 79. Are the **thermoluminescent materials** poor or good thermal conductors?

A. In general they are rather poor thermal conductors.

Q 80. How can the **temperature difference** between the two faces of a thermoluminescent chip be estimated?

A. The temperature difference can be estimated using the following expression

$$\Delta T = \frac{\beta \cdot t^2 \cdot c \cdot \rho}{k}$$

where

β = heating rate
t = sample thickness
c = specific heat`
ρ = density
k = thermal conductivity

Q 81. What are the factors affecting the **sensitivity** of a thermoluminescent instrument?

A. Voltage variations, PM tube fatigue, electronic instability, extremities in ambient weather, deposition of dirt and vapor on optical parts.

Q 82. Which factors can affect the **glow curve** of a thermoluminescent material?

A. The factors to take into consideration to get the same glow curves after similar irradiations are:

- reproducibility of the annealing procedures, i.e. pre- and post-iradiation annealing
- reproducibility of the red out cycle
- storage of the irradiated samples in constant and controlled environmental conditions
- reproducible cooling rate of the samples after the annealing procedure
- to avoid high irradiation doses if the samples are normally used in the linear region of response
- to avoid the exposure to sunlight, artificial light and sources of radiation
- use a cleaning procedure if contamination of samples is suspected
- regular check of the PM tube background using the reference light source included in the TL reader
- use vacuum tweezers when handling the samples
- take into account the elapsed time between irradiation and read out for fading corrections

Q 83. How do we define the **superlinearity** in the plot TL vs Dose?

A. The superlinearity is defined as the increase of the derivative of the dose dependence function.

Q 84. Is there any expression for a quantitative measure of the **superlinearity**?

A. Yes. It is the superlinearity index, defined by Chen and McKeever, given by the following expression:

$$g(D) = \left[\frac{D \cdot S''(D)}{S'(D)}\right] + 1$$

where $S'(D)$ is the first derivative of the dose dependence function, i.e. the TL signal vs dose D, and $S''(D)$ the second derivative.

Q 85. What kind of indication does the **superlinearity** index give?

A. It gives an indication of the change in the slope of the dose response.

Q 86. What is the **supralinearity** index $f(D)$?

A. It gives the amount of deviation from linearity in the TL vs Dose function.

Q 87. How is the **supralinearity** index $f(D)$ defined?

A. It is defined according to the following expression:

$$f(D) = \left[\frac{S(D) - S_0}{D}\right] \Big/ \left[\frac{S(D_l) - S_0}{D_l}\right]$$

where D_l is the normalization dose in the linear range of the TL vs D response and S_0 is the intercept on the TL axis of the extrapolated linear region.

Q 88. What is the physical reason that the area under a thermoluminescent **glow peak** is proportional to the dose absorbed by the TL dosimeter?

A. The trapped charge concentration, n, is evaluated from the area under the peak according to the following equation

$$n = \frac{1}{\beta} \int_{T_i}^{T_f} I \cdot dT'$$

where β is the linear heating rate, T_i and T_f are the temperatures at the beginning of the peak and at its end respectively, and I is the TL intensity. In turn, n is proportional to the absorbed dose D through the calibration factor of the system, F_C, i.e. $n = D/F_C$. So, we get

$$D = \frac{F_C}{\beta} \int_{T_i}^{T_f} I \cdot dT'.$$

Q 89. What is the meaning of **physicochemical stability** for a dosimetric material?

A. It means that a material used for dosimetric purposes should not undergo any physicochemical changes during repeated use, i.e. annealing process, repeated exposures, read out cycles.

Q 90. What is **tribothermoluminescence**?

A. This phenomenon is produced by friction during heating, giving some spurious signals. A large triboluminescence can be observed when monochristalline powder is used.

Q 91. Which are the characteristics of a **first order peak**?

A. According to the following figure

- The first order peaks are asymmetrical and $\tau = T_M - T_1$, the half-width at the low temperature side of the peak, is almost

50% larger than $\delta = T_2 - T_M$, the half-width towards the fall-off of the glow-peak $\tau \sim 1.5\delta$. The shape and the peak temperature depend on the heating rate.

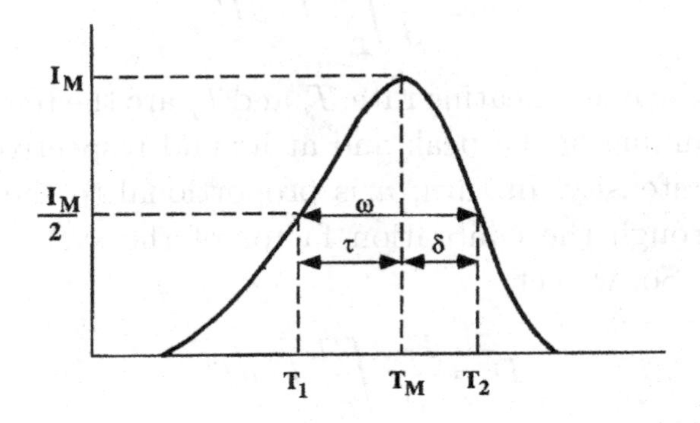

- The low temperature side of the peak, is almost 50% larger than $\delta = T_2 - T_M$, the half-width towards the fall-off of the glow-peak $\tau \sim 1.5\delta$. The shape and the peak temperature depend on the heating rate.
- For a fixed heating rate, both peak temperature and shape are independent of the initial trapped electron concentration n_0, as can be observed from the condition at the maximum

$$\frac{\beta E}{kT_M^2} = s \exp\left(-\frac{E}{kT_M}\right).$$

- The value of n_0 depends on the pre-measurement dose.
- The TL glow-curve obtained for any n_0 value can be superimposed onto the curve obtained for a different n_0 by multiplying by an appropriate factor.
- A first order peak is characterized by a geometrical factor $\mu = \delta/\omega = (T_2 - T_M)/(T_2 - T_1)$ equal to about 0.423.
- For fixed values of dose and heating rate, the ω value increases as E decreases.
- The decay at constant temperature of a first order peak is exponential.

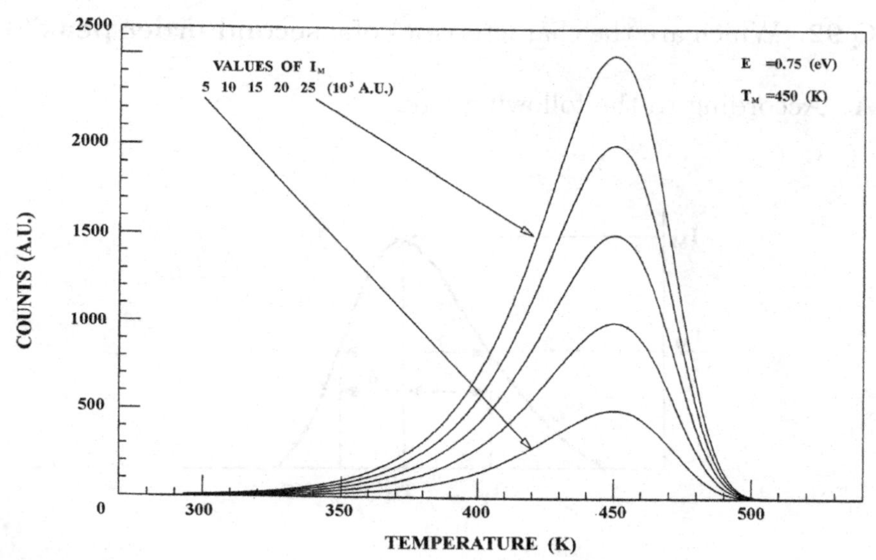

A computed first order glow-peak showing the linear increase of I_{M} as a function of dose.

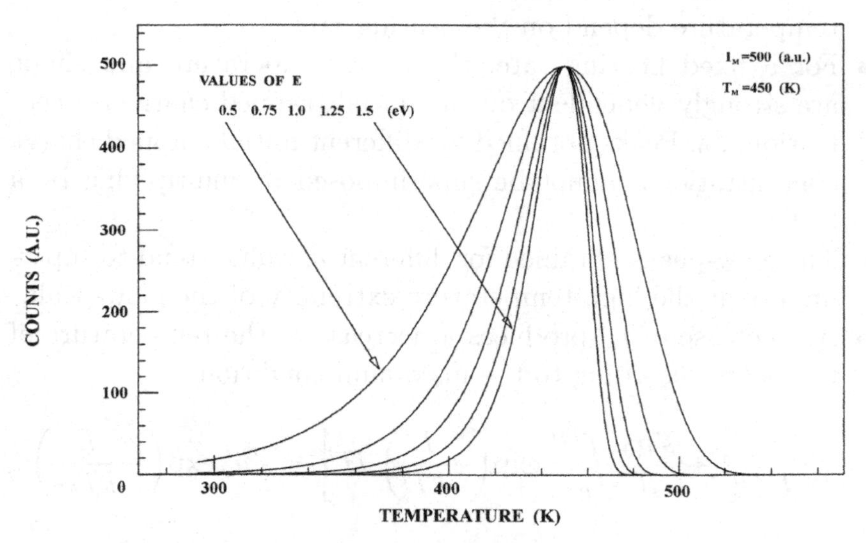

A computed first order glow-peak showing the increase of ω as E decreases.

Q 92. Which are the characteristics of a **second order peak**?

A. According to the following figure

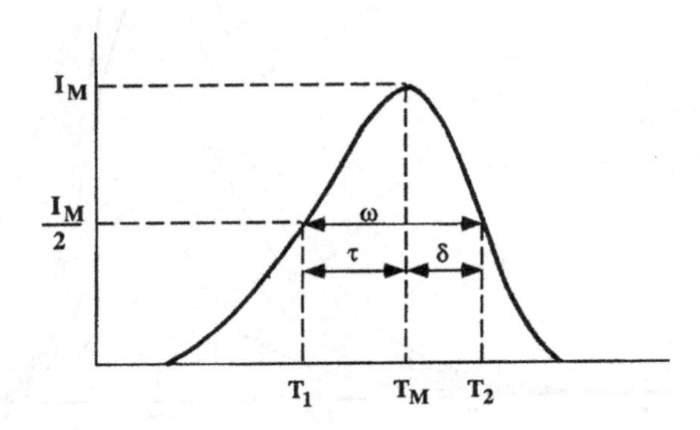

- A second order peak is practically symmetrical ($\delta \approx \tau$).
- To keep all other parameters constant, the shape and the peak temperature depend on the heating rate.
- For a fixed heating rate, the peak temperature and shape are strongly dependent on the initial trapped charge concentration, n_0. Peaks obtained for different initial trapped charge concentrations cannot be superimposed by multiplying by a factor.
- The glow-peaks obtained for different n_0 values tend to superimpose at the high temperature extremity of the glow-peak.
- An increase of n_0 produces a decrease in the temperature of the peak, according to the maximum condition

$$\frac{\beta E}{2kT_M^2}\left[1+\frac{s'n_0}{\beta}\int_{T_0}^{T_M}\exp\left(-\frac{E}{kT'}\right)dT'\right]=s'n_0\exp\left(-\frac{E}{kT_M}\right).$$

- The isothermal decay of a second order peak is hyperbolic.
- A second order peak is characterized by a geometrical factor $\mu \cong 0.524$.

- Furthermore, a decrease in the temperature of the peak, T_M, is observed as a function of the kinetics order changing from 1 to 2. This effect is illustrated in the figure below.

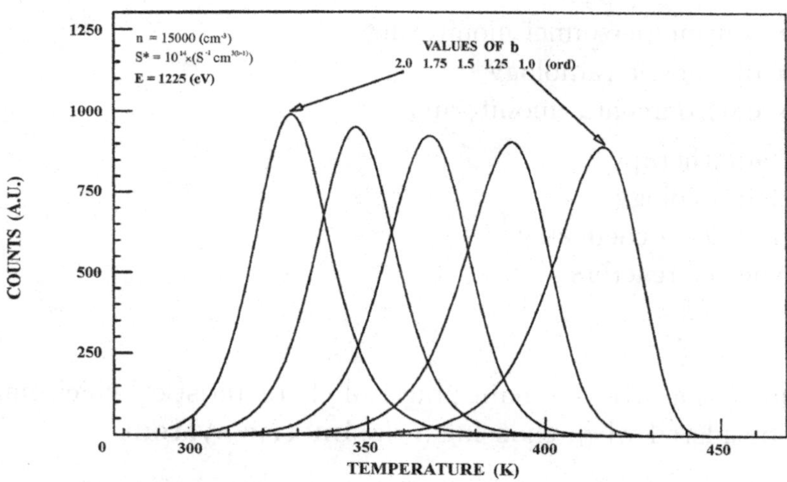

Computerized glow-peaks showing the effect of the kinetics order on the position of the peak temperature.

Q 93. What are the procedures if **contamination** of the thermoluminescent dosimeters is suspected?

A. A standard cleaning procedure is as follows:

a. Ultrasonic treatment in 1% detergent solution for 2 minutes.
b. Rinsing in running water for 2 minutes.
c. Rinsing in distilled water for 0.5 minutes.
d. Rinsing in alcohol for 0.5 minutes.
e. Drying of TLDs in a dust-free atmosphere.

Q 94. Which are the main applications for the **integrating dosimeters**?

A. The main applications of integrating dosimeters are summarized in the following:

1. Radiological protection
   - routine personnel monitoring
   - diagnostic radiology
   - environmental monitoring
2. Radiotherapy
3. Radiobiology
4. Radiation chemistry
5. Nuclear reactors

Q 95. What are the main practical characteristics which must be considered when choosing a **dosimetric system**?

A. The main practical characteristics to be considered are the following:

- the system should be able to measure the integrated dose
- to be able to distinguish between different types of radiation
- to have a large linear relationship between dosimeter response and dose
- to have a good accuracy according to the application
- to have a good precision of repeated measurements with a single dosimeter and with a batch of dosimeters
- the dosimeter should respond to radiation in the same manner as the medium in which the dose is to be measured
- the radiation field strength should not change throughout the volume of the dosimeter

Q 96. To which quantities is the **emitted light**, L(λ), from a dosimeter proportional to?

A. The emitted quantity of light $L(\lambda)$, due to an absorbed dose, detected and measured during the readout cycle will be proportional to:

D_0 the absorbed dose
$P(\lambda)$ the light generated in the dosimeter per unit absorbed dose
K_1 the geometrical-optical factor summarizing the effects of dosimeter shape and transparency, the dose distribution throughout the dosimeter and the effective geometrical efficiency of light collection
K_2 the fraction of the total light sum measured

Q 97. To which quantities is the **light generated** in the dosimeter **per unit absorbed dose**, P(λ), proportional to?

A. $P(\lambda)$ is a complex function depending on:

- the mean phosphor sensitivity
- the mean absorbed dose
- the thermal and radiation history of the dosimeter
- the time interval between irradiation and read out
- the heating rate during read out

Q 98. To which quantities is the **geometrical-optical factor**, K_1, a function of?

A. K_1 is a function of:

- μ the mass absorption coefficient of the radiation
- $A(\lambda)$ the absorption coefficient of the dosimeter for light generated within the dosimeter
- G the geometrical efficiency of light collection

- r the effective tray reflectivity
- the orientation of the dosimeter in the readout instrument

Q 99. To which parameters is the fraction of the **total light sum** measured, K_2, a function of?

A. The factor K_2 is a function of the heating cycle used and the interval over which the light is measured.

Q 100. How do we express analytically the **light $L(\lambda)$ emitted** by an irradiated dosimeter and reaching the PM tube of the reader?

A. $L(\lambda)$ is analytically expressed by the following relation:

$$L(\lambda) = P(\lambda) \cdot K_1 \cdot K_2 \cdot D_0$$

Q 101. How can be the **dose distribution** through a dosimeter simulated?

A. As it is illustrated in the following figure

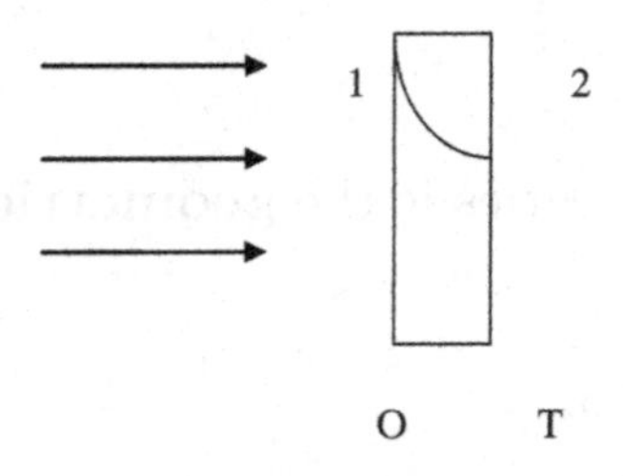

The dose distribution through a dosimeter, having a thickness T, is given by

$$D_X = D_0 \exp(-\mu \cdot x)$$

where μ is the mass energy absorption coefficient, x the distance from surface 1 within the dosimeter, and D_X is the dose deposited at a distance X from surface 1.

Q 102. How do we simulate the **self-attenuation of light** output by the dosimeter?

A. The light generated within a dosimeter will be attenuated in its path to the PM tube by absorption and scattering effects within the dosimeter. The attenuation is of the form

$$I_X = I_0 \exp[-A(\lambda) \cdot X]$$

where, according to the following figure

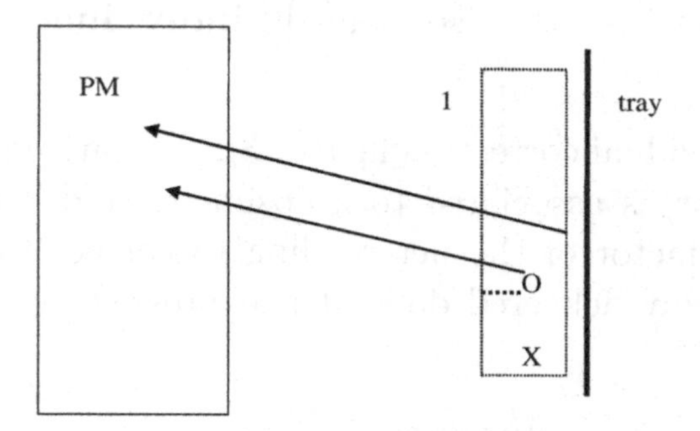

I_0 is the intensity of the light at point O, I_X is the intensity at a distance X from O, $A(\lambda)$ is the coefficient of attenuation by the dosimeter of its internally generated light.

Q 103. What is the **photon energy response** curve?

A. The photon energy response curve is the plot of the relative response of the thermoluminescent material to that of a reference material, i.e. air or tissue against photon energy for materials exposed to X or gamma rays. When the reference material is tissue, this plot gives an indication of the tissue equivalence of the material.

Q 104. What is the behavior of the **photon energy response** curve as a function of the photon energy?

A. At low photon energies (below about 150 keV) photoelectric absorption is the dominant interaction process with a cross section for interaction which increases approximately as Z^4, where Z is the effective atomic number of the absorbing material.

Q 105. What is the so called **individual correction factor**, S_i?

A. The individual correction factor, S_i, or relative intrinsic sensitivity factor, is associated to a dosimeter and it is used as a multiplying factor of the net reading to correct the dosimeter response at any delivered dose. It is expressed by the following expression:

$$M_{i,net(cor)} = M_{i,net} \cdot S_i$$

Q 106. What is the **relative intrinsic sensitivity factor**?

A. Another name for the individual correction factor.

Q 107. What is the practical meaning of the **individual correction factor**, S_i?

A. The S_i factor is a correction factor which is needed to avoid any reading variations owing to the individual sensitivity of each dosimeter.

Q 108. How to calculate the **individual correction factor**?

A. Considering a batch of N dosimeter, the individual correction factor is given by the following expression:

$$S_i = \frac{\bar{M}}{M_i - M_{0i}}$$

where

M_i is the reading of the i-th dosimeter
M_{0i} is the background reading of the i-th dosimeter
$M_i - M_{0i} = M_{i,net}$
$\bar{M}$ is the average of the net readings of the N dosimeters

Q 109. Which are the factors that affect the dosimeter **background**?

A. The factors that affect the dosimeter background are several and they can be listed as following:

- transparency variability of the dosimeters
- exposure to sun light
- exposure to artificial light
- dusty contamination of the dosimeter surface
- ineffective and non-reproducible pre- and/or post-irradiation annealing procedures
- changes in sensitivity due to radiation damage
- non-reproducible thermal read-out cycle
- self-dose effect if the thermoluminescent material includes some radioactive elements

Q 110. Can the thermoluminescent dosimeters show any **self-dose** effect?

A. Yes. The self-dose effect arises from a possible radioactive content of the thermoluminescent materials.

Q 111. How do we carry out an estimation of the **self-dose** effect, if any?

A. Suspect self-dose effect can be estimated leaving the annealed dosimeters in a sufficiently thick lead shield of about 5 cm to stop most of the external radiation.

Q 112. What is a **thermally disconnected trap**?

A. It is a trap which is very deep in the band gap so that during the sample heating, the trapped electrons are not released.

Q 113. What is the **correction factor**?

A. The correction factor is a numerical value which should be evaluated when a thermoluminescent material has a TL response at photon energies below 100 keV significantly greater than that at higher energies, i.e. Cs and Co.

Q 114. How do we calculate the **correction factor**?

A. The correction factor is obtained as the ratio between the calibration factor obtained with a reference beam quality, i.e. Cs or Co, and the calibration factor obtained with another beam quality.

Chapter 2

THEORY OF THERMOLUMINESCENCE

Q 115. What is the **activation energy**?

A. The activation energy is the energy associated with a metastable level within the forbidden gap of a crystal. It is expressed in eV.

Q 116. What is the **Arrhenius' equation**?

A. The equation of Arrhenius gives the mean lifetime that an electron spends in a trap at a given temperature. It is

$$\tau = s^{-1} \exp\left(\frac{E}{kT}\right).$$

Q 117. How do we describe the **absorption of light** in a solid?

A. The absorption of light in a solid is described by the Lambert-Beer law.

Q 118. What is the **Lambert-Beer law**?

A. It is the law describing the absorption of light in a solid.

According to this law, the absorption of light in a solid is described by the following equation:

$$I(\lambda, x) = I_0(\lambda) \exp[-\alpha(\lambda)x]$$

where

$I_0(\lambda)$ is the intensity of the incident light having a wavelength λ
$\alpha(\lambda)$ is the absorption coefficient
x the distance in the solid
$I(\lambda, x)$ the light intensity at position x

Q 119. What is the meaning of **GOT model**?

A. It means **General One Trap model**.

Q 120. Which of the different **thermoluminescent models** is the simplest?

A. The simplest model for thermoluminescence consists of two localized levels, the conduction and the valence bands CB and CV respectively, an isolated electron trap T and a recombination center, RC, as it is shown in the figure. This model is commonly referred to as the **One-Trap-One-Recombination center model** (**OTOR**).

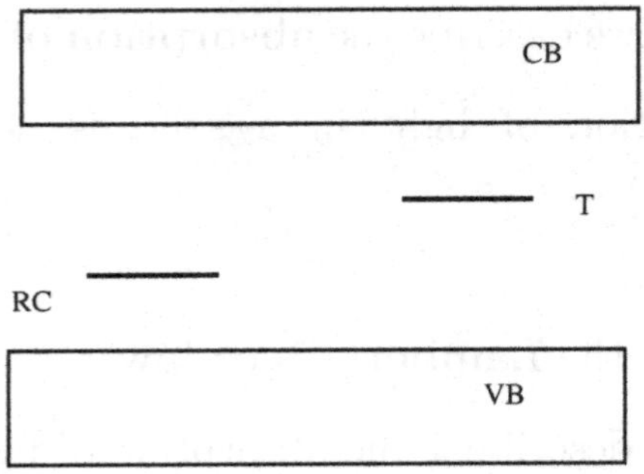

CB

T

RC

VB

Q 121. What is the meaning of **OTOR**?

A. It is the name of a TL model and it means One-Trap-One-Recombination center.

Q 122. Are the **GOT** and the **OTOR** models different?

A. No. They are just different names for the same model.

Q 123. What is the **OTOR model**?

A. The OTOR model consists of two localized levels, the conduction and the valence bands CB and CV respectively, one isolated electron trap, T, and a recombination center, RC.

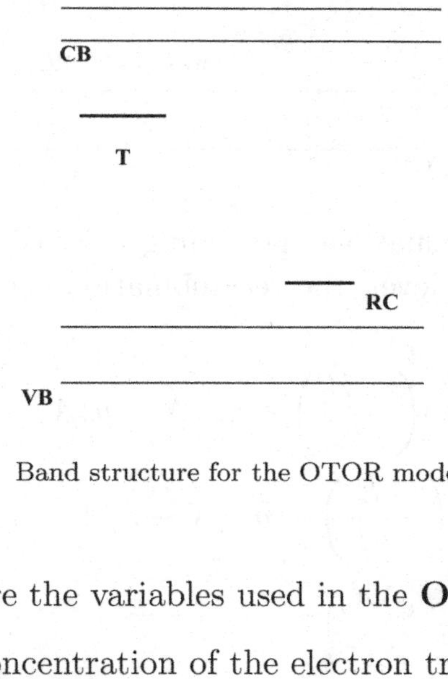

CB

T

RC

VB

Band structure for the OTOR model.

Q 124. What are the variables used in the **OTOR model**?

A. N = total concentration of the electron traps in the crystal (cm^{-3}), n = concentration of the filled electron traps in the crystal (cm^{-3}), n_C = concentration of the free carriers in the conduction band CB (cm^{-3}), E = activation energy of the electron traps (eV), s = frequency factor of the electron trap (s^{-1}), A_n = capture coefficient of the traps ($cm^3\ s^{-1}$), A_h = capture coefficient of the recombination center RC ($cm^3\ s^{-1}$).

Q 125. What are the **differential equations** in the **OTOR** model?

A. According to the following figure

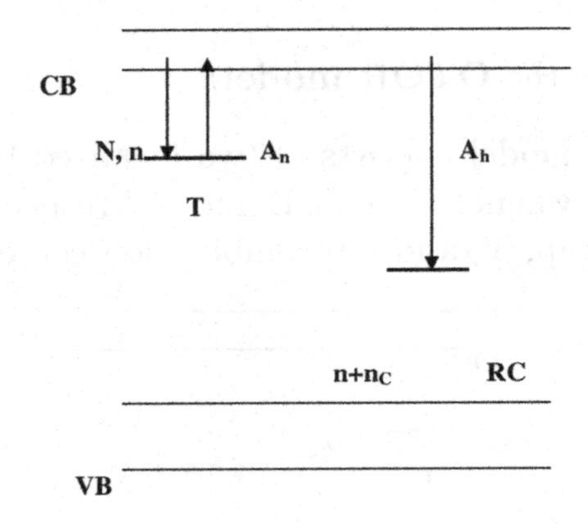

the differential equations governing the traffic of electrons between the trap level, the recombination center and the conduction band are:

$$\frac{dn}{dt} = -ns\exp\left(-\frac{E}{kT}\right) + n_C(N-n)A_n$$

$$\frac{dn_C}{dt} = ns\exp\left(-\frac{E}{kT}\right) - n_C(N-n)A_n - n_C(n+n_C)A_h$$

$$TL = n_c(n+n_C)A_h$$

Q 126. What is the meaning of the quantity $\frac{dn}{dt}$ in the differential equations of the **OTOR model**?

A. $\frac{dn}{dt}$ represents the rate of change of concentration of electrons $n(t)$ in the trap as the sample is heated during the thermoluminescence measurement.

Q 127. What is the meaning of the quantity $\frac{dn_C}{dt}$ in the differential equations of the **OTOR model**?

A. $\frac{dn_C}{dt}$ represents the rate of change of concentration of electrons in the conduction band during heating of the thermoluminescent sample.

Q 128. What is the term describing the **thermal excitation** of the electrons leaving the trap?

A. The thermal excitation is described mathematically by the term

$$ns \exp\left(-\frac{E}{kT}\right).$$

Q 129. What is the physical meaning of the differential equations governing the **OTOR model**?

A. The first equation describes the traffic of electrons in and out of the electron traps. The electrons can leave the traps via thermal excitation, which is described mathematically by the term $ns \exp\left(-\frac{E}{kT}\right)$. The electrons can also be retrapped in the trap: this event is described by the so-called retrapping term $n_C(N - n)A_n$.

The second equation describes the traffic of electrons in and out of the conduction band. The electrons in the conduction band can be trapped in the recombination center RC. This event is described by the term $n_C(n + n_C)A_h$.

The third equation gives the thermoluminescent signal, which is proportional to the amount of light measured during the thermoluminescence measurement.

The quantity $n + n_C$ is the total concentration of the filled traps at any moment. Because of conservation of charge, this quantity is also equal to the total concentration of filled holes in the recombination center.

Q 130. How do we solve the differential equations of the **OTOR** model?

A. The differential equations of the OTOR model can be solved only numerically.

Q 131. Which cases are described by the **OTOR** model?

A. The OTOR model allows to describe the first, the second and the general order of kinetics.

Q 132. Which is the equation giving the **maximum condition** for a first order glow peak?

A. The maximum condition of a first order glow peak is given by the following equation:

$$\frac{\beta E}{kT_M^2} = s \exp\left(-\frac{E}{kT_M}\right)$$

where β is the linear heating rate (°C/s), E is the activation energy (eV), k the Boltzmann's constant, T_M the absolute temperature (K) and s the frequency factor (s^{-1}).

Q 133. Which is the equation giving the **maximum condition** for a second order glow peak?

A. The maximum condition of a second order glow peak is given by the following equation:

$$\frac{\beta E}{2kT_M^2}\left[1+\frac{s'n_0}{\beta}\int_{T_0}^{T_M}\exp\left(-\frac{E}{kT'}\right)dT'\right]=s'n_0\exp\left(-\frac{E}{kT_M}\right)$$

where β is the linear heating rate (°C/s), E is the activation energy (eV), k the Boltzmann's constant, T_M the absolute temperature (K) at the maximum, s' the pre-exponential factor ($\mathrm{cm}^3\ \mathrm{s}^{-1}$) and n_0 is the ccentration of the initial number of the trapped charges (cm^{-3}).

Q 134. Which is the equation giving the **maximum condition** for a general order glow peak?

A. The maximum condition of a general order glow peak is given by the following equation:

$$\frac{kT_M^2 bs}{\beta E}\exp\left(-\frac{E}{kT_M}\right)=1+\frac{s(b-1)}{\beta}\int_{T_0}^{T_M}\exp\left(-\frac{E}{kT'}\right)dT'$$

where β is the linear heating rate (°C/s), E is the activation energy (eV), k the Boltzmann's constant, T_M the absolute temperature (K) at the maximum, s' the pre-exponential factor ($\mathrm{cm}^3\ \mathrm{s}^{-1}$), n_0 is the ccentration of the initial number of the trapped charges (cm^{-3}) and b is the kinetics order.

Q 135. What is the **tunneling** phenomenon?

A. When a trap and a recombination center are spatially close to each other, it is possible to have recombination without any involvement of the delocalized bands.

Q 136. Why can the **initial rise method** be expressed by the equation $I(T) \propto \exp(-\frac{E}{kT})$?

A. This is because the amount of trapped electrons, n, in the low temperature tail of a TL peak can assumed to be constant since the dependence of n on temperature is negligible in that temperature region.

Q 137. What is the **interactive model**?

A. This model is used when charges released from one trap, in general a shallow trap, can be retrapped in a deeper trap before undergoing light-emitting recombination. The deep trap is called **thermally disconnected** trap. The model is also called IMTS.

Q 138. What is the meaning of **IMTS**?

A. IMTS means Interactive Multi Trap System.

Q 139. What is the band model for the **interactive thermoluminescence** process?

A. The band model is shown in the figure below.

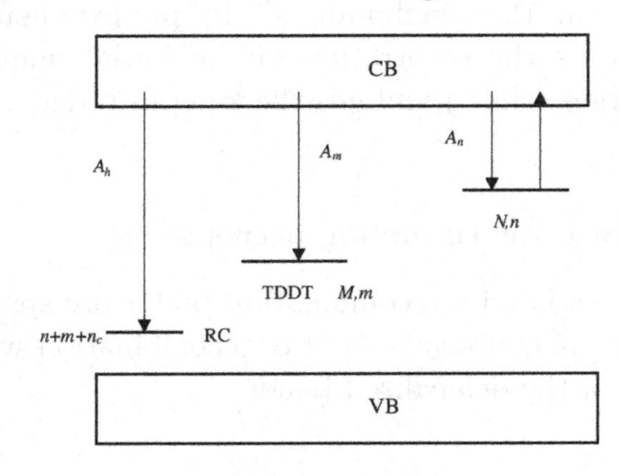

List of variables used in the model:

RC = recombination center
TDDT = thermally disconnected deep trap
N = total concentration of the electron traps (cm^{-3})
M = total concentration of TDDT electron traps
n = concentration of the filled electron traps (cm^{-3})
n_c = concentration of the free charges in CB (cm^{-3})
m = concentration of the filled TDDT electron traps (cm^{-3})
E = activation energy of the electron traps (eV)
s = frequency factor of the electron traps (s^{-1})
A_n = capture coefficient of the traps (cm^{-3} s^{-1})
A_h = capture coefficient of the recombination center RC (cm^{-3} s^{-1})
A_m = capture coefficient of the TDDT (cm^{-3} s^{-1})

Q 140. Which equations describe the **interactive thermoluminescent** process?

A. The interactive thermoluminescent process is described by the following differential equations:

$$\frac{dn}{dt} = -ns\exp\left(-\frac{E}{kT}\right) + n_C(N-n)A_n$$

$$\frac{dm}{dt} = n_C(M-m)A_m$$

$$\frac{dn_C}{dt} = ns\exp\left(-\frac{E}{kT}\right) - n_C(N-n)A_n - n_C(m-M)A_m$$
$$- n_C(m+n+n_C)A_h$$

$$TL = n_C(m+n+n_C)A_h$$

Q 141. What is the physical meaning of the equations governing the **interactive thermoluminescent** process?

A. The first equation describes the traffic of electrons in and out of the electron traps. The electrons can leave the traps via thermal excitation, described by $ns\exp\left(-\frac{E}{kT}\right)$. The electrons can also be retrapped in the trap and this is described by the retrapping term $n_C(N-n)A_n$. The second equation describes the traffic of electrons from the conduction band into TDDT; this event is given by the term $n_C(M-m)A_m$. The third equation describes the traffic of electrons in and out of the conduction band. The electrons in the conduction band can be trapped in the recombination center RC and this event is described by $n_C(m+n+n_C)A_h$. The forth equation gives the TL signal, which is proportional to the amount of light measured during the thermoluminescence measurement.

The quantity $(m+n+n_C)$ is equal to the total concentration of filled traps at any moment. Because of conservation of charge, this quantity is also equal to the total concentration of filled holes in the recombination center.

The previous set of equations can be solved only by using numerical method.

Q 142. What is the **assumption required for first and second order TL kinetics**?

A. The first and second order kinetics require that separate processes control each glow peak. It is assumed that charges released from one type of trap do not interact with other type of traps.

Q 143. Is there a general description of thermoluminescence from which it is possible to obtain the various **kinetics orders**?

A. E. I. A. Adirovitch, in 1956, used a set of three differential equations to explain the decay of phosphorescence in the general case. The same model was then applied by Haering-Adams (1960) and Halperin-Braner (1960) to describe the flow of charges between localized energy levels and delocalized bands during trap emptying. The solution of the differential equations allows us to obtain both first and second order kinetics equations.

Q 144. Why does the **Adirovitch model** not allow us to obtain the general order kinetics?

A. Because the general order equation is an empirical expression introduced by May and Partridge for taking into account experimental situations having intermediate kinetics processes.

Q 145. Is there any experimental evidence for a **kinetics order** less than one?

A. Yes. Partridge and May (1965) explained their experimental results introducing two competing processes: a radiative process following a first order and a zero order without radiative transitions.

Q 146. What is the common assumption leading to **kinetics orders** larger than one?

A. The assumption is that the concentration of recombination centers is equal to the number of trapped electrons.

Q 147. What is the **Adirovitch** theory?

A. Adirovitch used a set of three differential equations to explain the decay of phosphorescence.

Q 148. What is the energy level diagram used by **Adirovitch**?

A. It is given in the following figure:

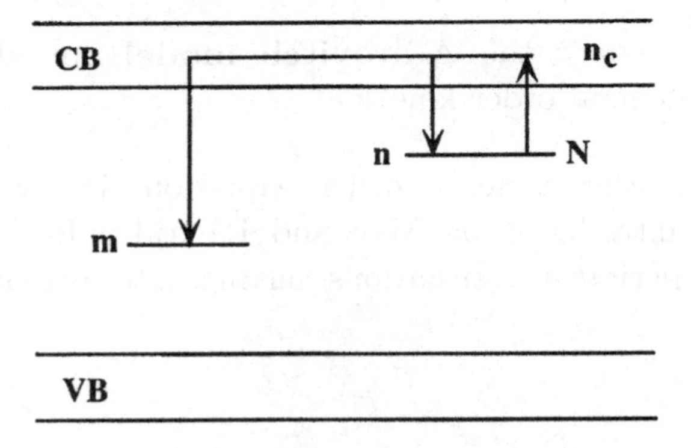

Q 149. What are the equations of **Adirovitch**?

A. The equations of Adirovitch are the following. The first equation is related to the emission intensity:

$$I = -\frac{dm}{dt} = A_m m n_c$$

where

I is the emission intensity
m is the concentration of recombination centers (holes in centers) (cm^{-3});

n_c is the concentration of free electrons in the conduction band (cm^{-3});
A_m is the recombination probability ($\mathrm{cm}^3\ \mathrm{s}^{-1}$).

The second equation deals with the population variation of electrons in traps, n (cm^{-3}), and it takes into account the excitation of electrons into the conduction band as well as the possible retrapping. Then we have:

$$\frac{dn}{dt} = -sn \exp\left(-\frac{E}{kT}\right) + n_c(N-n)A_n$$

where A_n($\mathrm{cm}^3\ \mathrm{s}^{-1}$) is the retrapping probability and N (cm^{-3}) is the total concentration of traps.

The third equation relates to the charge neutrality. It can be expressed as

$$\frac{dn_c}{dt} = \frac{dm}{dt} - \frac{dn}{dt}.$$

Q 150. Can the **Adirovitch** theory be extended to the thermoluminescence process?

A. Yes. The equations of Adirovitch were applied to thermoluminescence by Halperin and Braner.

Q 151. What are **intrinsic defects**?

A. The intrinsic defects, also called native defects, can be of the following:

- vacancies or missing atoms, called Schottky defects: one atom is extracted from its site and not replaced
- interstitial or Frenkel defect: one atom of the crystal is in a non-proper lattice site

- substitutional defects: i.e. halides ions in alkali sites
- aggregation of the previous defects

Q 152. Can **irradiation** produce **defects** in a crystal?

A. Yes. As an example it is possible to have:

- F center: one electron trapped in a positive vacancy
- V_K center: a hole is trapped by a pair of negative ions

Q 153. Which the **extrinsic** or **impurity defects** are?

A. The extrinsic defects, also called impurity defects, are generated by chemical elements which are inserted in a crystal during its growth, i.e. Cu in LiF.

Q 154. Who introduced the **first order kinetics** model?

A. The first order kinetics was introduced by **Randall and Wilkins** in 1945.

Q 155. What are the assumptions of the **first order kinetics**?

A. The assumptions are:

- Irradiation of the TL sample at a low enough temperature so that no electrons are released from the traps
- The lifetime of the electrons in the conduction band is short
- The luminescence efficiency of the recombination centers is temperature independent
- The concentrations of traps and recombination centers are temperature independent
- No electrons released from a trap is retrapped

Q 156. What is the band model used by Randall and Wilkins to describe the **first order kinetics process**?

A. The band model used by Randall and Wilkins to describe a first order process is given in the following figure:

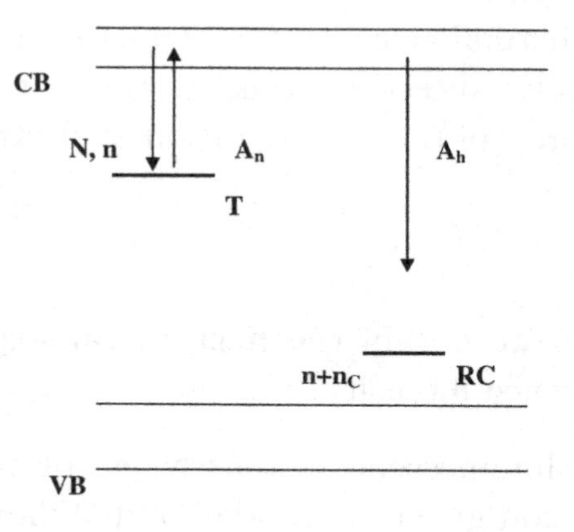

Q 157. What is the meaning of the variables used in the **Randall and Wilkins model**?

A. N = total concentration of the electron traps in the crystal (cm^{-3}), n = concentration of the filled electron traps in the crystal (cm^{-3}), n_C = concentration of the free carriers in the conduction band CB (cm^{-3}), E = activation energy of the electron traps (eV), s = frequency factor of the electron trap (s^{-1}), A_n = capture coefficient of the traps (cm^3 s^{-1}), A_h = capture coefficient of the recombination center RC (cm^3 s^{-1}).

Q 158. What is the **first order** equation?

A. The first order equation is the following:

$$\frac{dn}{dt} = -ns\exp\left(-\frac{E}{kT}\right).$$

Q 159. What is the meaning of the terms comparing in the **first order** equation?

A. $\frac{dn}{dt}$ represents the rate of change of concentration of electrons $n(t)$ as the sample is heated during the thermoluminescence measurement. The term $ns \exp(-\frac{E}{kT})$ describes mathematically the thermal release of electrons from the traps. The observed thermoluminescent intensity, I_{TL}, is equal to the negative rate of change of the concentration of electrons in the trap, i.e. $I_{TL} = -\frac{dn}{dt}$.

Q 160. How do we obtain the **first order** expression for the thermoluminescence intensity?

A. The thermoluminescence intensity is obtained solving the first order equation given by Randall and Wilkins.

Q 161. What is the expression for the thermoluminescence intensity in the case of **first order** kinetics?

A. Assuming a linear rate of heating, β, the first order differential equation can be solved to yield the following solution:

$$I(T) = sn_0 \exp\left(-\frac{E}{kT}\right) \exp\left[-\frac{s}{\beta}\int_{T_0}^{T} \exp\left(-\frac{E}{kT'}\right) dT' + 1\right]$$

where n_0 is the initial concentration of filled traps at time $t = 0$ and k is the Boltzmann constant.

Q 162. Who introduced the **second order kinetics** model?

A. The second order kinetics was introduced by Garlick and Gibson in 1948.

Q 163. Which are the assumptions of the **second order kinetics**?

A. The assumptions for a second order kinetics are the same as the first order kinetics, except for the possibility of retrapping: i.e. the probability of retrapping is large, so that the dominating process in a second order kinetics is retrapping.

Q 164. What is the band model used by Garlick and Gibson to describe the **second order kinetics process**?

A. The band model used by Garlick and Gibson to describe a second order process is given in the following figure:

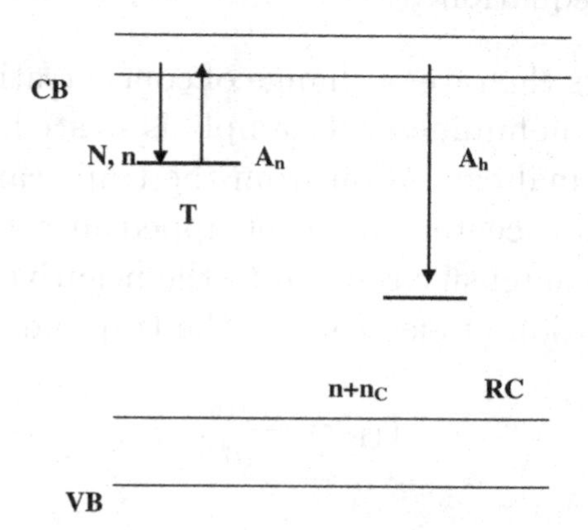

Q 165. What is the meaning of the variables used in the **Garlick and Gibson model?**

A. N = total concentration of the electron traps in the crystal (cm^{-3}), n = concentration of the filled electron traps in the crystal (cm^{-3}), n_C = concentration of the free carriers in the conduction band CB (cm^{-3}), E = activation energy of the electron traps (eV), s = frequency factor of the electron trap (s^{-1}), A_n = capture coefficient of the traps (cm^3 s^{-1}),

A_h = capture coefficient of the recombination center RC (cm^3 s^{-1}).

Q 166. What is the **second order** kinetics equation?

A. The second order kinetics equation is the following:

$$\frac{dn}{dt} = -\frac{n^2}{N} s \exp\left(-\frac{E}{kT}\right).$$

Q 167. What is the meaning of the terms comparing in the **second order** equation?

A. $\frac{dn}{dt}$ represents the rate of change of concentration of electrons $n(t)$ as the thermoluminescent sample is heated. The electrons released by thermal excitation from the traps can recombine in the recombination center or be retrapped in a trap. The thermoluminescence intensity is equal to the negative rate of change of the concentration of electrons in the trap, i.e.

$$I_{\mathrm{TL}} = -\frac{dn}{dt}.$$

Q 168. How to obtain the **second order** expression for the thermoluminescence intensity?

A. The thermoluminescence intensity is obtained solving the second order equation given by Garlik and Gibson.

Q 169. What is the expression for the thermoluminescence intensity in the case of **second order** kinetics?

A. Assuming a linear heating rate, β, the second order differential equation can be solved to yield the following solution:

$$I(T) = \frac{s}{N} n_0^2 \exp\left(-\frac{E}{kT}\right)\left[1 + \frac{s n_0}{\beta N}\int_{T_0}^{T} \exp\left(-\frac{E}{kT'}\right) dT'\right]^{-2}.$$

Q 170. Who introduced the **general order kinetics** model?

A. This model was introduced by May and Partridge in 1964.

Q 171. What is the band model used by May and Partridge to describe the **general order kinetics process**?

A. The band model used by **May and Partridge** to describe a **general order** process is given in the following figure:

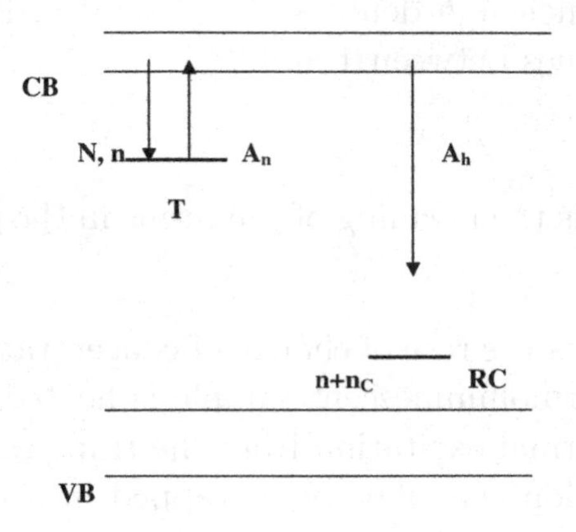

Q 172. In which cases can a **general order** kinetics be used?

A. When experimental situations indicate intermediate kinetics processes, between 1 and 2.

Q 173. What is the meaning of the variables used in the **May and Partridge model**?

A. N = total concentration of the electron traps in the crystal (cm^{-3}), n = concentration of the filled electron traps in the crystal (cm^{-3}), n_C = concentration of the free carriers in the conduction band CB (cm^{-3}), E = activation energy of the electron traps (eV), s = frequency factor of the electron trap (s^{-1}), A_n = capture coefficient of the traps ($cm^3\ s^{-1}$), A_h = capture coefficient of the recombination center RC ($cm^3\ s^{-1}$).

Q 174. What is the **general order kinetics** equation?

A. The general order kinetics equation is the following:

$$\frac{dn}{dt} = n^b \frac{s}{N} \exp\left(-\frac{E}{kT}\right)$$

where the coefficient b denotes the general order kinetics and can assume values between 1 and 2.

Q 175. What is the meaning of the terms in the **general order** equation?

A. $\frac{dn}{dt}$ represents the rate of change of concentration of electrons $n(t)$ as the thermoluminescent sample is heated. The electrons released by thermal excitation from the traps can recombine in the recombination center or be retrapped in a trap. The thermoluminescence intensity is equal to the negative rate of change of the concentration of electrons in the trap, i.e.

$$I_{TL} = -\frac{dn}{dt}.$$

Q 176. How do we obtain the **general order expression** for the thermoluminescence intensity?

A. The thermoluminescence intensity is obtained by solving the general order equation given by May and Partridge.

Q 177. What is the expression for the thermoluminescence intensity in the case of **general order kinetics**?

A. Assuming a linear heating rate, β, the general order differential equation can be solved to yield the following solution:

$$I(T) = \frac{sn_0^b}{N} \exp\left(-\frac{E}{kT}\right) \times \left[1 + \frac{s(b-1)n_0^{b-1}}{\beta N} \int_{T_0}^{T} \exp\left(-\frac{E}{kT'}\right) dT'\right]^{-\frac{b}{b-1}}.$$

Q 178. Can the **general order** equation

$$\frac{dn}{dt} = n^b \frac{s}{N} \exp\left(-\frac{E}{kT}\right)$$

give the first and the second order kinetics?

A. Yes, for $b = 1$, the general order equation reduces to the first order kinetics and for $b = 2$, it reduces to the second order kinetics.

Q 179. There is any **thermoluminescent model** which considers that both electrons and holes may be released from their traps at the same time and in the same temperature interval?

A. Yes. This kind of model has been discussed by **Schön-Klasens**.

Q 180. Which is the **Schön-Klasens** model?

A. The Schön-Klasens model is depicted in the following figure:

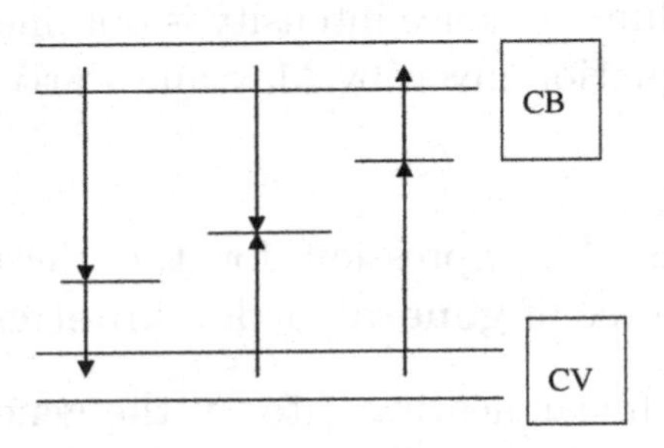

Schön-Klasens model.

It has to be noted that the holes are thermally released from the same centers which act as recombination sites for the thermally released electrons.

Q 181. What is happening if the trapped holes are released during **thermal stimulation**?

A. The release of trapped holes removes recombination centers for electrons.

Q 182. What is the consequence of canceling **recombination centers?**

A. A possible consequence is the thermal quenching of luminescence, as it was discussed by Schön in 1942.

Q 183. What are the rate equations of the **Schön-Klasens** model?

A. Referring to the following figure

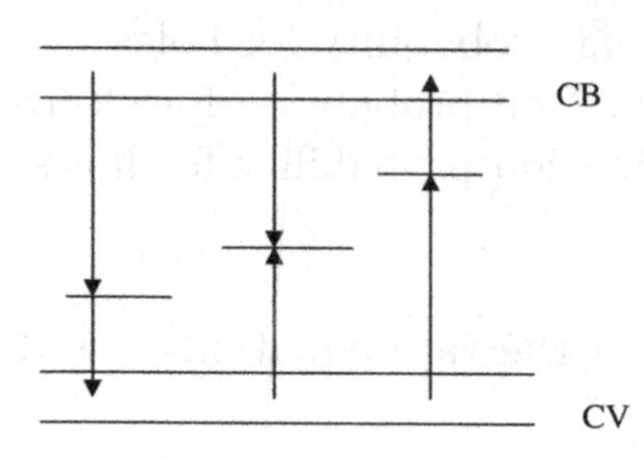

the rate equations describing the charge carrier transitions between the energy levels during thermal excitation are the following:

$$\frac{dn_c}{dt} = np_n - n_c(N - n)A_n - n_c n A_{mn}$$

$$\frac{dn_v}{dt} = mp_p - n_v(M - m)A_p - n_v n A_{np}$$

$$\frac{dn}{dt} = n_c(N - n)A_n - n_v n A_{np} - np_n$$

$$\frac{dm}{dt} = n_v(M - m)A_p - n_c m A_{mn} m p_p$$

where

n_c = concentration of free electrons in CB
n_v = concentration of free holes in CB
n = concentration of trapped electrons
m = concentration of hole states for recombination
N = trapping sites concentration for electrons
M = concentration of available hole traps
$N - n$ = concentration of empty electron traps
$M - m$ = concentration of empty hole traps
p_n = probability for thermal excitation of electrons from traps
p_p = probability for thermal excitation of holes from traps

A_n = retrapping probability for electrons
A_p = retrapping probability for holes
A_{mn} = recombination probability for electrons
A_{np} = recombination probability for holes

Q 184. Who solved the rate equations based on the model of **Schön-Klasens**?

A. Braunlich and Scharmann, in 1966, solved the rate equations based on the Schön-Klasens model, introducing several simplifying assumptions.

Q 185. What kind of assumptions were made by **Braunlich and Scharmann** in solving the rate equations based on the **Schön-Klasens** model?

A. Braunlich and Scharmann defined the following parameters:

$$R = \frac{A_n}{A_{mn}} \quad \text{and} \quad R' = \frac{A_p}{A_{np}}$$

expressing the ratios of the retrapping probabilities compared to the recombination probabilities.

They then solved the rate equations for four different cases:

- $R \prec\prec 1, \quad R' \prec\prec 1$
- $R \succ\succ 1, \quad R' \prec\prec 1$
- $R \prec\prec 1, \quad R' \succ\succ 1$
- $R \succ\succ 1, \quad R' \succ\succ 1$

They also used the assumptions $n \prec\prec N$, $m \prec\prec M$, the neutrality condition $n_c + n = n_{v+m}$ and the quasiequilibrium (or QE) assumption.

Q 186. Can the **recombination** take place without transition of electrons into the conduction band?

A. Yes. This kind of recombination is called localized transition and it is depicted in the following figure:

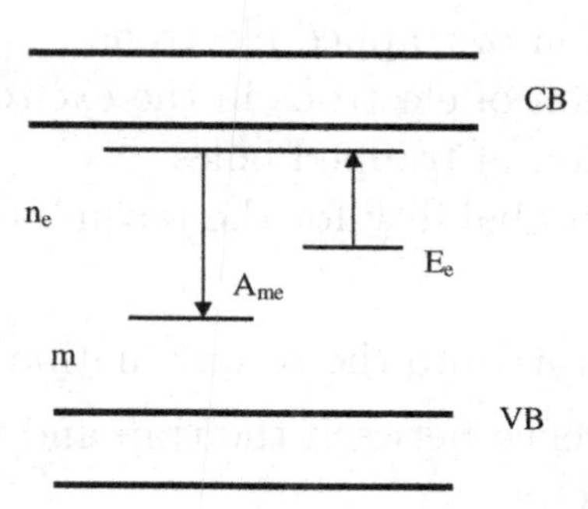

where

n_e = concentration of electrons in the excited state
m = concentration of trapped holes
A_{me} = transition probability for the recombination of an electron from the excited state into the recombination center
E_e = energy difference between the trap and the excited state

Q 187. Who mathematically described the **localized transitions**?

A. In 1960, **Halperin and Braner**, gave the following rate equations to describe this process:

$$I_{TL} = -\frac{dn}{dt} = n_e m A_{me}$$

$$\frac{dn}{dt} = sn_e - s\exp\left(-\frac{E_e}{kT}\right)$$

$$\frac{dn_e}{dt} = sn\exp\left(-\frac{E_e}{kT}\right) - n_e(mA_{me} - s)$$

where

I_{TL} = thermoluminescent intensity
n = concentration of trapped electrons
n_e = concentration of electrons in the excited state
m = concentration of trapped holes
A_{me} = transition probability for the recombination of an electron

from the excited state into the recombination center

E_e = energy difference between the trap and the excited state
s = frequency factor

Q 188. How to define the **isothermal decay** of the thermoluminescent signal?

A. The isothermal decay is the escape of electrons from traps at higher temperatures (held constant) than the ambient temperature.

Q 189. In what way is the **faded thermoluminescence** dependent on the trapped charges?

A. The faded thermoluminescence is directly proportional to n, the number of trapped electrons.

Q 190. In what way does the measured thermoluminescence intensity during an **isothermal decay** depend on the trapped charges?

A. The measured thermoluminescence intensity during an isothermal decay is directly proportional to the rate at which the detrapping occurs, i.e. it is proportional to dn/dt.

Q 191. What is the best way to describe the **fading** characteristics of a phosphor?

A. The fading characteristics can be best described by the lifetime, $\mathtt{t}$, of the particular trapping level.

Q 192. Is there any correlation between the lifetime of a trapping level and the **concentration of the activators** in a phosphor?

A. In most of the phosphors the optimum concentration of the activators is related to the maximum stability of the thermoluminescent signal, i.e. great lifetime. Furthermore, the optimum concentration of the activators gives also the maximum thermoluminescence yield.

Q 193. What is the difference between the **frequency factor** s and the **pre-exponential factor** s'?

A. The frequency factor s is a constant for a particular trap. When it is dependent on the initial concentration of the carriers in a trap, it is referred as the pre-exponential factor s'.

Q 194. Is there a relationship between the **frequency factor** and the **pre-exponential factor**?

A. The relation between the frequency factor and the pre-exponential factor is the following:

$$s = s'n_0$$

where n_0 is the initial concentration of the carriers in the trap.

Q 195. Is the **pre-exponential factor** depending on temperature?

A. In general the pre-exponential factor does not depend on temperature, but in certain cases it is found to be temperature dependent.

Q 196. In what way is the **pre-exponential factor** temperature dependent?

A. In general, the following relation is valid:

$$s' = s'_0 T^\alpha$$

where α has a value ranging from -2 to $+2$, depending on the recombination cross section as a function of temperature.

Q 197. Which expression gives the **probability rate of escape** of the carriers from a trap?

A. The probability rate of escape from a trap, p, is given by the following expressions

$$p = s\exp(-E/kT)$$

where s is the frequency factor, E the activation energy associated to the trap, k the Boltzmann constant and T the absolute temperature.

Q 198. Is there any relationship between the **probability rate of escape**, p, and the **lifetime**, t, of the charge carriers in the trap?

A. Yes. The lifetime, t, of the charge carriers in the trap at a given temperature T, is the inverse of the escape rate probability p, i.e.

$$\mathtt{t} = p^{-1} = s^{-1} \exp(E/kT).$$

Q 199. What is the physical meaning of the **frequency factor**?

A. Considering a trap as a potential well, the frequency factor represents the product of the number of times an electron hits the walls and the wall reflection coefficient.

Q 200. What is the order of magnitude of the **frequency factor**?

A. The order of magnitude of the frequency factor should be similar to the vibrational frequency of a crystal.

Q 201. What is the possible range of the values of the **frequency factor** encountered in thermoluminescent materials?

A. The values of the frequency factor are usually in the range of $10^8 - 10^{11}$ sec^{-1}.

Q 202. Is the **frequency factor** s dependent on temperature?

A. The frequency factor s may be considered in some cases to be dependent on temperature.

Q 203. What kind of relationship exists between the **frequency factor** and the temperature?

A. The frequency factor is proportional to a power of the absolute temperature.

Q 204. What is the mathematical relation between the **frequency factor** and the temperature?

A. The relation between the frequency factor and the temperature is the following:

$$s = s_0 T^{\alpha}$$

where α has various values in the range -2 and $+2$.

Q 205. What is the first order detrapping rate when the **frequency factor** is temperature dependent?

A. The first order detrapping rate when the frequency factor is temperature dependent is given by

$$\frac{dn}{dt} = -n s_0 T^{\alpha} \exp\left(-\frac{E}{kT}\right).$$

Q 206. What is the first order expression for the **thermoluminescence intensity**, I, when the **frequency factor** is temperature dependent?

A. In this case the TL intensity is given by the following expression:

$$I(T) = n_0 s_0 T^{\alpha} \exp\left(-\frac{E}{kT}\right)\left[-\frac{s_0}{\beta}\int_{T_0}^{T} T^{\alpha} \exp\left(-\frac{E}{kT'}\right) dT'\right].$$

Q 207. What is the temperature T^* at which the **electron escape probability** is 1 $\sec^{-1}$?

A. From

$$p = s \exp(-E/kT^*)$$

we get, for $p = 1$

$$E = kT^* \ln s$$

and then

$$T^* = E/k \ln s.$$

Q 208. Which are the criteria that should be checked for a **second order** peak?

A. The criteria to be observed are the following:

- A second order peak exhibits a geometrical factor $\mu = 0.52$
- The position of maximum TL intensity shifts toward higher temperature values for low doses
- A second order peak may exhibit superlinearity at low doses
- The isothermal decay for a second order peak shows a graph of $(\frac{I_t}{I_0})^{-1/2}$ versus time which should yield a straight line of slope E

Q 209. What are the criteria that should be checked for a peak following a **general-order** kinetics?

A. The criteria to be observed are the following:

- A peak following a general-order kinetics should have a geometrical factor μ having a value between 0.42 and 0.52
- The peak position shifts toward low temperature values as the given dose increases
- General order kinetics corresponds to decay curves described by a plot of $(\frac{I_t}{I_0})^{\frac{1-b}{b}}$ versus time. A straight line is obtained for a suitable value of b

Q 210. What is the possible error on the **geometrical factor** μ?

A. The standard deviation, SD, of μ is given by the following expression:

$$SD = \pm \sqrt{\begin{array}{l} \left(\frac{T_2 - T_M}{(T_2 - T_1)^2} \cdot \Delta T_1\right)^2 + \left(\frac{T_M - T_1}{(T_2 - T_1)^2} \cdot \Delta T_2\right)^2 \\ + \left(-\frac{1}{T_2 - T_1} \cdot \Delta T_M\right)^2 \end{array}}$$

where T_1, T_2 and T_M correspond to the half maximum intensity and to the maximum respectively.

Q 211. What is the physical significance of the nature of the **glow curve**?

A. When a luminescent material is excited by ionizing radiations, equal numbers of electrons and holes are liberated within it and some of them may become trapped at certain centers in the material, i.e. energy is stored in the material.

Q 212. What are the **storage sites** called?

A. They are called electron or hole traps, according to the nature of the electrical carriers being trapped.

Q 213. What is the meaning of the **peak temperature**?

A. The peak temperature is a measure of the thermal activation energy needed for the detrapping.

Q 214. What are the criteria to be fulfilled to be sure that a **glow peak** follows **first-order kinetics**?

A. The criteria to be checked are the following:

- A first-order peak has a geometrical factor μ equal to 0.42
- The position of the peak at the maximum does not shift in temperature for different radiation doses
- The isothermal decay of a first-order peak follows an exponential behavior

Q 215. What is the physical situation for **first order kinetics**?

A. The first order kinetics corresponds to a situation in which the probability for electron-hole recombination is very large and, in turn, the probability of retrapping is negligible.

Q 216. What is the physical situation for **second order kinetics**?

A. The second order kinetics corresponds to a situation in which a very large retrapping probability exists.

Q 217. How do we recognize at a first sight if a TL **glow peak** is following a **first** or a **second order process**?

A. A first order peak is asymmetrical and a second order peak is, on the contrary, practically symmetric.

Q 218. Who proposed the **two-trap model**?

A. The two trap model was proposed by Sweet and Urquhart (1980) to explain a situation where two TL peaks are so close that they appear as only one peak.

Q 219. How many processes are involved in **thermoluminescence** (TL) and in **optically stimulated luminescence** (OSL)?

A. Thermoluminescence and optically stimulated luminescence are two-stage processes: one is the excitation stage, the second is the read-out stage.

Q 220. In what way does the **excitation stage** and/or the **read-out stage** influence the TL and the OSL emissions?

A. The two stages can influence the growth of the signal versus dose: i.e. superlinear or supralinear growth, as well as alterations in the material sensitivity.

Q 221. Are there any models to explain **superlinearity**?

A. There are mainly two models; one considers competition among traps during excitation and the second, competition during read out.

Q 222. What is the energy levels model describing the **competition** during excitation?

A. It is given the following figure:

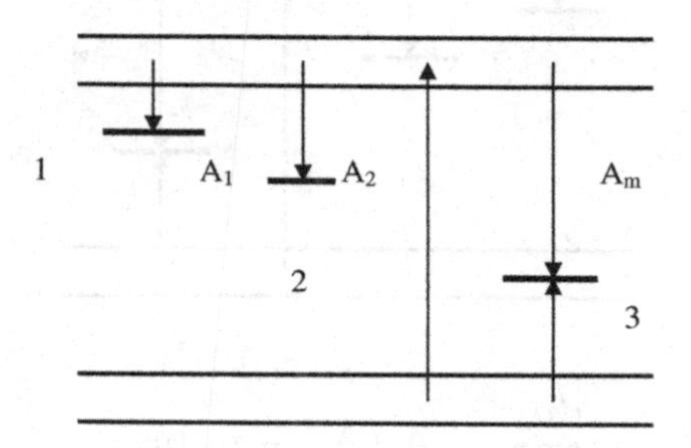

The meaning of the symbols is:

A_1-trapping probability of active trap
A_2-trapping probability of competing trap
A_m-recombination probability
1-active trap level of concentration N_1 and occupancy n_1
2-deep competing trap level of concentration N_2 and occupancy n_2
3-radiative recombination center of concentration M and occupancy m

Q 223. Who proposed the model concerning **competition** during excitation?

A. The model was initially proposed by Suntharalingam and Cameron in 1967; then it was modified by Aitken *et al.* (1968) and finally by Chen and Bowman (1978).

Q 224. What are the equations governing the **competition** during excitation?

A. The equations governing the electron transitions are given according to the following figure:

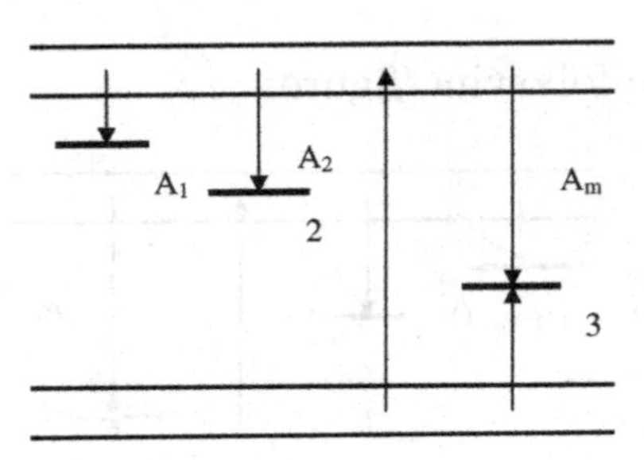

$$\frac{dn_1}{dt} = A_1(N_1 - n_1)n_c$$

$$\frac{dn_2}{dt} = A_2(N_2 - n_2)n_c$$

$$\frac{dn_c}{dt} = X - \frac{dn_1}{dt} - \frac{dn_2}{dt}$$

The meaning of the symbols is:

A_1-trapping probability of active trap
A_2-trapping probability of competing trap
A_m-recombination probability
1-active trap level of concentration N_1 and occupancy n_1
2-deep competing trap level of concentration N_2 and occupancy n_2
3-radiative recombination center of concentration M and occupancy m

n_c-concentration of electrons in CB
X-rate of creation of electron-hole pairs by irradiation.

Q 225. What is the energy levels model used to describe the **competition** during readout?

A. It is given in the following figure.

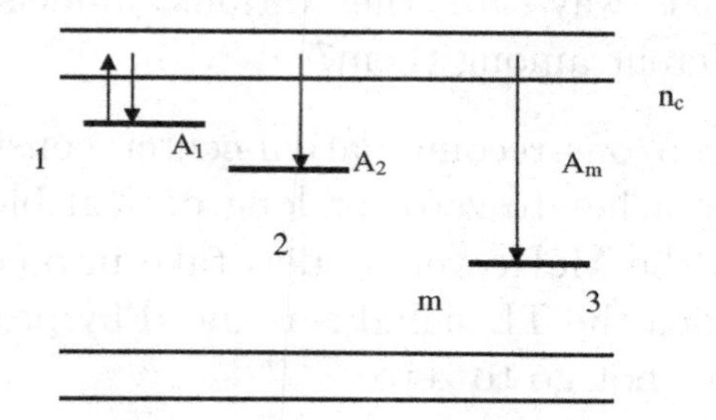

Q 226. What are the equations governing the **competition** during readout?

A. The equations governing competition during readout, according to the previous figure, are the following (Kristianpoller *et al.*, 1974)

$$\frac{dn_1}{dt} = -sn_1 \exp\left(-\frac{E_1}{kT}\right) + A_1 n_c (N_1 - n)$$

$$\frac{dn_2}{dt} = A_2 (N_2 - n_2) n_c$$

$$I_{TL} = -\frac{dm}{dt} = A_m m n_c$$

$$\frac{dm}{dt} = \frac{dn_1}{dt} + \frac{dn_2}{dt} + \frac{dn_c}{dt}$$

Q 227. Is there any model to explain the **optical bleaching** of thermoluminescence?

A. Yes. There are mainly three models: one is based on the simple one-trap/one-recombination-center; two more have been proposed by R. Chen (1990) and by S. W. S. McKeever (1991).

Q 228. In what way are the various models for **optical bleaching** different among them?

A. The one-trap/one-recombination-center considers that the TL signal approaches to zero for long optical bleaching times. The Chen and the McKeever models take into consideration a situation in which the TL signal is reduced by prolonged illumination, but does not go to zero.

Q 229. According to the model of one-trap/one-recombination-center (see figure below), what are the equations for **optical bleaching**?

A. The equations, assuming the one-trap/one-recombination-center, are the following:

- the detrapping rate is given by

$$\frac{dn}{dt} = -nf + n_c(N - n)A_n$$

- the optical excitation rate is given by

$$f(\lambda) = \Phi(\lambda)\sigma_0(\lambda)$$

where Φ is the photon fluence, σ_0 is the optical cross-section, n is the concentration of the trapped charges, n_c is the concentration of free electrons in the CB, N is the concentration of available traps, and A_n is the retrapping probability.

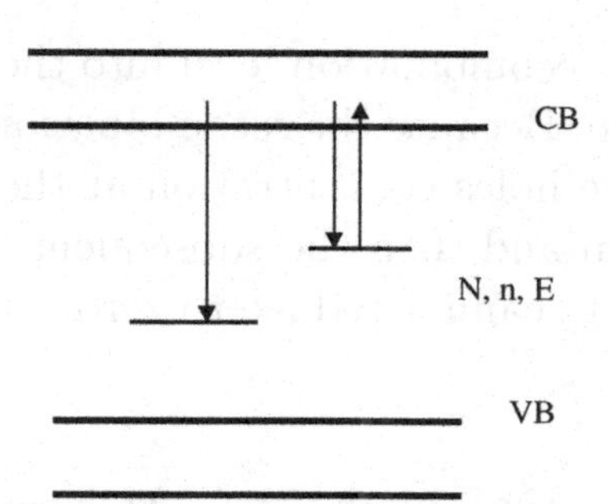

Q 230. What is the optical equation of the **GOT model**?

A. The optical equation of the GOT model is

$$\frac{dn}{dt} = -\frac{nfmA_m}{(N-n)A_n + mA_m}.$$

Q 231. What is the model of Chen for **optical bleaching**?

A. The model of Chen is given in the following figure

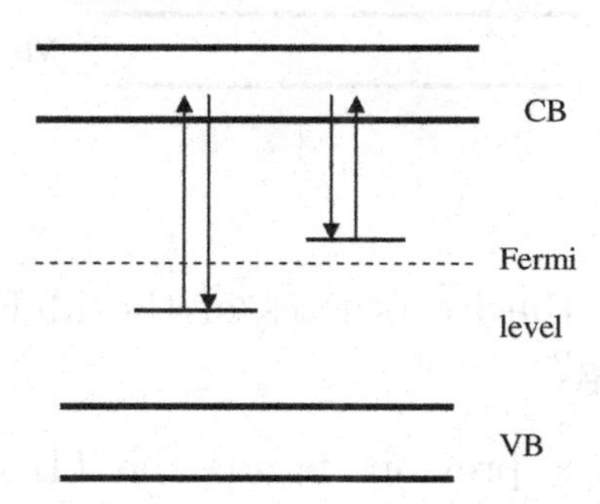

Q 232. Which are the hypotheses of Chen's model for **optical bleaching**?

A. The model allows for retrapping into the electron trap; furthermore, the model also considers the optical excitation of electrons from the recombination level into the conduction band during illumination. Because the recombination centers lie below the Fermi level, the holes concentration at the centers increases during illumination and then the subsequent TL signal (during heating for readout) cannot reduce to zero.

Q 233. What is the model of McKeever for **optical bleaching**?

A. The model is shown in the figure below:

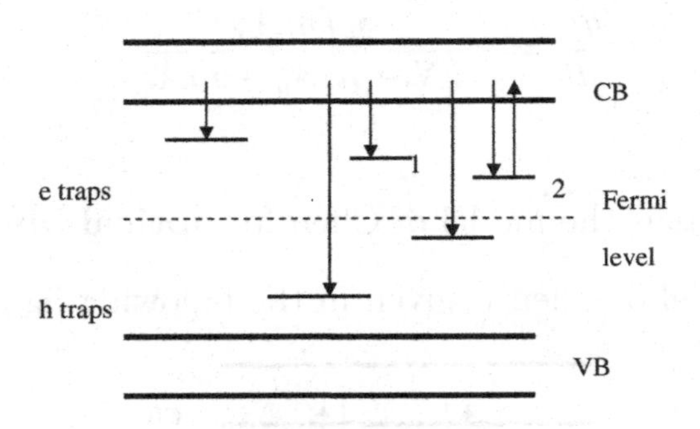

Q 234. What are the hypotheses of the McKeever model for **optical bleaching**?

A. According to the previous figure, the TL signal is given by the release of electrons from level 1, but the illumination does not bleach this level because it empties charges from the deep level 2 instead of from level 1 which, in turn, does not go to zero.

Q 235. What is the simplest model for **phototransferred** thermoluminescence (**PTTL**)?

A. The simplest model for phototransferred thermoluminescence takes into consideration a shallow trap, a deep trap and a recombination center.

Q 236. What are the functions of the shallow and deep traps in the model of **phototransferred** thermoluminescence?

A. The shallow trap is the one where the charges are transferred from the deep trap.

Q 237. Is there a mathematical description for the **phototransfer** phenomena?

A. The mathematical model for the phototransfer is given by the following rate equations:

$$\frac{dn_2}{dt} = -n_2 f + n_c(N_2 - n_2)A_{n2}$$

$$\frac{dn_1}{dt} = n_c(N_1 - n_1)A_{n1}$$

$$\frac{dm}{dt} = -n_c m A_m$$

where

n_1, n_2 are the concentration of electrons in the shallow and deep traps respectively
m is the concentration of the holes in the recombination centers
n_c is the concentration of electrons in the CB
N_1 and N_2 are the concentrations of the shallow and deep traps respectively
A_{n1} and A_{n2} are the retrapping probabilities for the shallow and deep traps respectively
A_m is the recombination probability

f is the rate of excitation of electrons from deep traps during illumination

Q 238. Who suggested an approximated expression for the **integral** $\int_{T_0}^{T} \exp\left(-\frac{E}{kT'}\right) dT'$ comparing in the thermoluminescence theory?

A. The approximated expression was given by C. H. Haake (1975).

Q 239. What kind of approximation has to be used for the **integral** $\int_{T_0}^{T} \exp\left(-\frac{E}{kT'}\right) dT'$ comparing in the thermoluminescence theory?

A. The approximation is given by the asymptotic series

$$\int_{T_0}^{T} \exp\left(-\frac{E}{kT'}\right) dT' \cong T \exp\left(-\frac{E}{kT}\right) \sum_{1}^{n} \left(\frac{kT}{E}\right)^{n} (-1)^{n-1} n!$$

Q 240. How many terms of the **asymptotic series** are necessary for a good approximation of the integral?

A. It is correct to use only the first two terms of the series:

$$\int_{T_0}^{T} \exp\left(-\frac{E}{kT'}\right) dT' \cong T \frac{kT}{E} \exp\left(-\frac{E}{kT}\right) - T_0 \frac{kT_0}{E} \exp\left(-\frac{E}{kT_0}\right)$$

Q 241. How do we explain the observed **thermoluminescent properties** of a solid?

A. An explanation can be obtained from the energy band theory of solids.

Q 242. How does the **energy band model** of solids work for thermoluminescence?

A. In a real crystal, semiconductor or insulator material, some energy levels are located in the forbidden gap between the conduction and the valence bands. In a simple TL model two levels are assumed, one situated below the bottom of the conduction band and the other situated above the top of the valence band, as it is shown in the figure. The highest level, T, is above the equilibrium Fermi level and it is empty in the equilibrium state. It acts as a potential electron trap. The second level, RC, is below the Fermi level and it is a possible hole trap and acts as a recombination center.

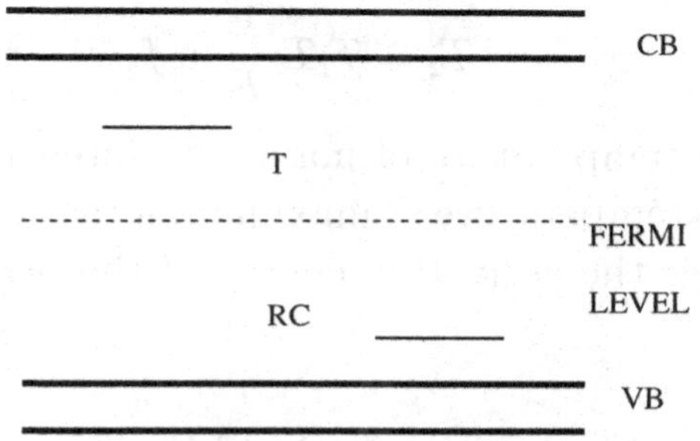

CB

T

FERMI LEVEL

RC

VB

Q 243. What is the expression of the **TL intensity** for the one trap-one recombination center?

A. The equation, as a function of time, is

$$I(t) = \frac{mA_m ns \exp\left(-\frac{E}{kT}\right)}{(N-n)A_n + mA_m}$$

where

m is the concentration of the trapped holes in RC (m^{-3})
n is the concentration of trapped electrons in T (m^{-3})
N is the concentration of the electron traps (m^{-3})

A_n is the retrapping probability (m^3 s^{-1})
A_m is the recombination probability (m^3 s^{-1})
s is the frequency factor (s^{-1})
E is the trap depth or activation energy (eV)
k is the Boltzmann's constant ($= 8.617 \times 10^{-5}$ eV/K)
T is the absolute temperature (K)

Q 244. Is it possible to estimate the **shift in temperature** for a second order peak as a function of the delivered dose?

A. Bos and Dielhof (1991) gave an approximated expression for the temperature shift in a second order peak. The expression is

$$T_1 - T_2 \cong T_1 T_2 \frac{k}{E} \ln f$$

where T_1 is the temperature of maximum intensity at a certain dose and T_2 the temperature of maximum intensity at an f times higher dose. E is the activation energy of the peak.

Q 245. Is it possible to estimate the **shift in temperature** for a general order peak as a function of the delivered dose?

A. Bos and Dielhof (1991) give an approximated expression for the temperature shift in a general order peak. The expression is

$$T_1 - T_2 \cong T_1 T_2 \frac{k(b-1)}{E} \ln f$$

where T_1 is the temperature of maximum intensity at a certain dose and T_2 the temperature of maximum intensity at an f times higher dose, E is the activation energy of the peak and b is the kinetics order.

Q 246. Is there any practical expression which gives the **TL intensity** for a first order peak?

A. Kitis *et al.* (1998) gave a quite accurate expression for the TL intensity for a first order peak:

$$I(T) = I_M \exp\left[1 + \frac{E}{kT}\frac{T - T_M}{T_M} - \frac{T^2}{T_M^2} \times \exp\left(\frac{E}{kT}\frac{T - T_M}{T_M}\right)\left(1 - \frac{2kT}{E}\right) - \frac{2kT_M}{E}\right]$$

Q 247. Is there any practical expression which gives the **TL intensity** for a second order peak?

A. Kitis *et al.* (1998) gave a quite accurate expression for the TL intensity for a second order peak:

$$I(T) = 4I_M \exp\left(\frac{E}{kT}\frac{T - T_M}{T_M}\right) \times \left[\frac{T^2}{T_M^2}\left(1 - \frac{2kT}{E}\right)\exp\left(\frac{E}{kT}\frac{T - T_M}{T_M}\right) + 1 + \frac{2kT_M}{E}\right]$$

Q 248. How to define the intrinsic **thermoluminescence efficiency**?

A. The intrinsic thermoluminescence efficiency is defined as the ratio between the energy emitted as visible light during heating of a TL sample and the energy absorbed during the exposure to ionizing radiation.

Q 249. How much is the typical **thermoluminescent efficiency** induced by ionizing radiation?

A. Typical values of thermoluminescence efficiency range from 0.01 to 1%.

Q 250. There is any expression giving the intrinsic **thermoluminescence efficiency**?

A. The intrinsic thermoluminescence efficiency is given by the following expression:

$$\eta = \frac{hv}{cE_g} \eta_{tr} p S Q \eta_{esc}$$

where

η_{tr} is the fraction of charge carriers, captured in traps, which can be thermally stimulated
η_{esc} is the fraction of produced photons that will escape without being absorbed by the TL sample
hv is the average energy of the emitted photons for TL emission
E_g is the width of the forbidden gap
c is a numerical value depending of the type of thermoluminescent material ranging between 1 and 4
p is the probability of release of charges from the traps
S is the efficiency of luminescent center
Q is the quantum efficiency of the luminescent center

Q 251. There is any theoretical formulation giving the **general order** kinetics equation of an isolated thermoluminescent peak where the pre-exponential factor results are independent of the dose of irradiation?

A. M. S. Rasheedy suggested a new thermoluminescence general order equation which seems to overcome the variation of the pre-exponential factor as a function of the dose.

Q 252. What is the new **general order** kinetics equation introduced by Rasheedy?

A. The new rate equation is the following

$$\frac{dn}{dt} = -\frac{n^b}{N^{b-1}} s \exp\left(-\frac{E}{kT}\right)$$

where the frequency factor s has units s^{-1}, similar to the frequency factor in the first kinetics order.

n and N are the trapped electrons concentration and the traps concentration respectively.

Q 253. What is the expression for the **trapped electrons** as obtained from the Rasheedy's equation?

A.

$$n = \frac{n_0}{\left[1 + \frac{s(b-1)\left(\frac{n_0}{N}\right)^{b-1}}{\beta} \cdot \int_{T_0}^{T} \exp\left(-\frac{E}{kT'}\right) dT'\right]^{\frac{1}{b-1}}}$$

where n_0 is the initial concentration of electrons in the traps.

Q 254. What is the equation for the **thermoluminescence intensity** according to the Rasheedy's equation?

A. The thermoluminescence intensity is given by

$$I = \frac{n_0^b s \exp\left(-\frac{E}{kT}\right)}{N^{b-1}\left[1 + \frac{s(b-1)\left(\frac{n_0}{N}\right)^{b-1}}{\beta} \cdot \int_{T_0}^{T} \exp\left(-\frac{E}{kT'}\right) dT'\right]^{\frac{b}{b-1}}}.$$

Q 255. What is the equation for the **fluorescence** decay?

A. The decay equation for the fluorescence is

$$I = I_0 \exp(-\alpha \cdot t)$$

where

α is the decay constant of the process, equal to $1/\tau$ with τ the lifetime of the excited state of the center
I_0 is the initial luminescence intensity just when the excitation is ceased ($t = 0$)

Q 256. What is the main characteristic of the decay equation of the **fluorescence**?

A. The equation indicates a monomolecular or first order decay kinetics.

Q 257. How do we express the **lifetime** of the free carriers?

A. The lifetime of the free carriers is given by the following equation:

$$\tau = \frac{1}{\rho \cdot \sigma \cdot u}$$

where

ρ is the number of recombination centers per cm^3
σ is the capture cross section
u is the thermal velocity of a carrier

Q 258. What is the meaning of the **quasi-equilibrium** (QE) assumption?

A. The quasiequilibrium assumption states that the free electron concentration in the conduction band is quasistationary.

Q 259. How do we express analytically the **quasi-equilibrium** (QE) assumption?

A. The QE assumption is expressed by

$$\left|\frac{dn_c}{dt}\right| \prec\prec \left|\frac{dn}{dt}\right|, \left|\frac{dm}{dt}\right|$$

where

n_c is the concentration of free electrons in the conduction band
n is the concentration of trapped electrons
m is the concentration of available hole states for recombination

Q 260. Why is the **quasi-equilibrium** (QE) assumption very important?

A. The QE assumption allows a simplification of the rate equations.

Q 261. Is there any other meaning of the **QE assumption**?

A. If the QE assumption is combined with the additional hypothesis that the initial free carrier concentration is small, $n_{co} \approx 0$, it means that free charge never accumulates in the conduction band during thermal stimulation.

Q 262. Can the **cavity theory** for photons be applied to the thermoluminescent dosimeters?

A. A **cavity theory** applied to TLDs has been proposed by T. E. Burlin in 1966.

Q 263. What are the assumptions of Burlin for the **cavity theory** applied to TLDs?

A. Burlin proposed the cavity theory for photons including a gamma generated source term in the cavity as well as perturbation of the medium generated electron spectrum in the cavity.

Q 264. How is the **half-life** of a trap defined?

A. The half-life of a trap, at a constant temperature, is defined as the time for which the number of trapped electrons fall to half of their initial value.

Q 265. What is the **half-life** expression for a first order process?

A. The half-life for a first order process is expressed by the following relation

$$t_{1/2} = \frac{0.693}{s \exp\left(-\frac{E}{kT}\right)}.$$

Q 266. What is the effect on the **half-life** of changing the activation energy value and keeping the temperature as a constant and for a given value of the frequency factor?

A. The half-life, at a constant temperature and for a given value of the frequency factor, increases as the activation energy increases. This is a logical consequence to the fact that the higher the activation energy, the higher the stability of the trap.

Q 267. Is there any expression for the **half-life** in the case of the second order kinetics process?

A. The half-life in this case is given by the following equation:

$$t_{1/2} = \frac{1}{n_0 s' \exp\left(-\frac{E}{kT}\right)}.$$

Q 268. What is the difference between the **half-life** for a first order kinetics and the one for the second order?

A. The half-life in the case of a first order kinetics is independent of the initial concentration of the trapped charges. On the other hand, the half-life for a second order kinetics is inversely proportional to the initial concentration of the trapped charges.

Q 269. What is the consequence on the **half-life** for a second order kinetics which is inversely proportional to the initial concentration of the trapped charges?

A. Because the half-life of a second order kinetics is dependent on the concentration of the trapped charges, i.e. the dose, the half-life is not a decreasing value as a function of the elapsed time from the initial irradiation, as in the first order case: the half-life increases as the initial concentration decreases.

Q 270. Is the increase of the **half-life** as a function of time only related to a second order kinetics?

A. No. The half-life increases too in the case of a general order kinetics.

Q 271. What is the effect of introducing **defects** and **impurities** in a crystal?

A. The introduction of defects and impurities in a crystal creates states within the energy gap of the host material. Because these states can absorb the light from some part of the visible spectrum, the crystal is colored and the states are termed color centers. These centers are also termed carrier traps.

Q 272. How is the **luminescence efficiency** defined?

A. The luminescence efficiency of a sample is defined by the ratio between the total energy emitted in the form of light and the absorbed energy during the process of its excitation.

Q 273. What is the meaning of **radiative transition**?

A. The radiative transition happens when an electron recombines in a recombination center with the emission of light, i.e. thermoluminescent glow curve is observed.

Q 274. What is the meaning of **non-radiative transition**?

A. In this process the energy of the electron escaped from the trap is transferred to another electron or to the crystal lattice (multiphonon processes) and then no light is emitted.

Chapter 3

KINETICS METHODS

Q 275. Are there any **approximate methods** for calculating the activation energy by just using the peak temperature at the maximum?

A. Yes. Randall-Wilkins and Urbach used the following expressions

$$E = 25kT_M$$

$$E = \frac{T_M}{500}$$

The two previous equations have been calculated using frequency factor values $2.9 \cdot 10^9\,\text{sec}^{-1}$ and $10^9\,\text{sec}^{-1}$ respectively.

Q 276. What is the **Initial Rise** method?

A. The Initial Rise (IR) method allows us to calculate the activation energy of a TL peak without any knowledge of the frequency factors.

Q 277. What is the mathematical expression used for the **Initial Rise** method?

A. The mathematical expression for the Initial Rise is:

$$I(T) \propto \exp\left(-\frac{E}{kT}\right).$$

Q 278. How does the **Initial Rise** method work?

A. According to the following figure, only the first part of the ascending peak is used, because in that part the TL intensity, $I(T)$ is proportional to $\exp(-\frac{E}{kT})$.

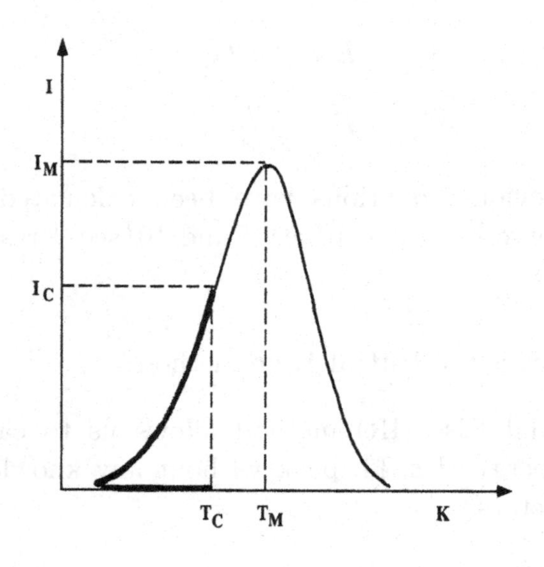

Q 279. What is a typical plot of the **Initial Rise method?**

A. It is shown in the figure

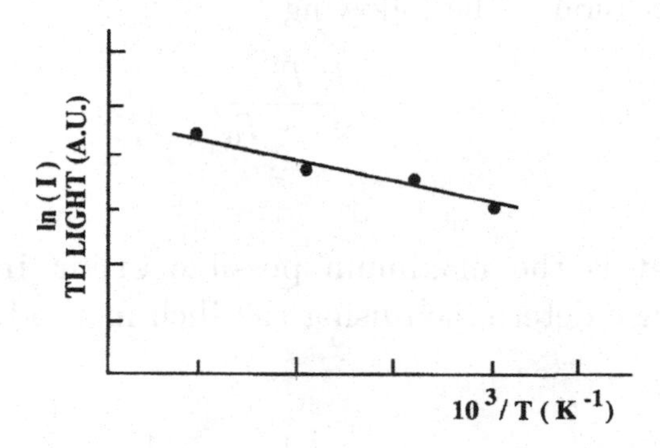

Q 280. Other than the Initial Rise method, is there any other method based on the low temperature tail of a TL peak in determining the activation energy?

A. Yes, there is the **Ilich's method**.

Q 281. How does the Ilich's method work?

A. In the following figure the method is depicted

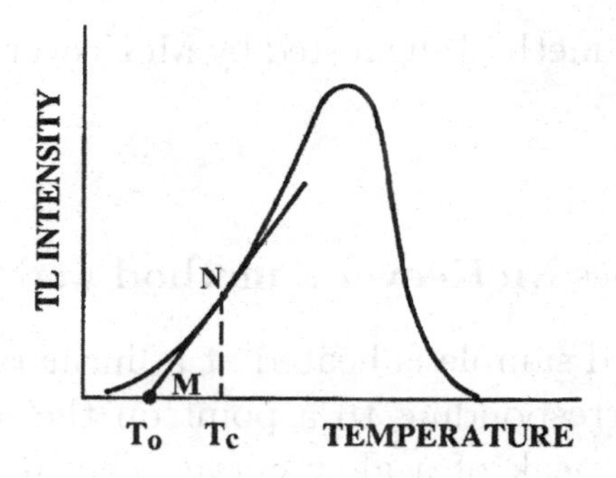

Q 282. What is the expression for the activation energy determined by the **Ilich's method**?

A. The expression is the following

$$E = k\frac{T_c^2}{T_c - T_0}.$$

Q 283. What is the maximum possible **error in the activation energy** determined using the Ilich method?

A.

$$\Delta E = 2 \cdot \Delta T \cdot E \left| \frac{1}{T_c} + \frac{1}{T_c - T_0} \right|.$$

Q 284. Is the two **heating rates method** independent from the frequency factor?

A. Yes.

Q 285. There is any method which enables us to separate **overlapped peaks** in the glow curve?

A. Yes. It is the method suggested by McKeever, i.e. the $T_{MAX}-T_{STOP}$ method.

Q 286. How does **McKeever's method** work?

A. An irradiated sample is heated at a linear rate to a temperature T_{STOP} corresponding to a point on the low temperature side of the first peak of a glow curve. The maximum temperature of the peak, T_{MAX} is noted and the sample is cooled down.

The procedure is repeated several times using irradiated samples increasing the value of T_{STOP}. After that a plot of T_{MAX} vs T_{STOP} is done.

Q 287. What is the **method of Sweet-Urquhart**?

A. The method is similar to the one of McKeever. It allows us to separate two overlapping glow peaks.

Q 288. What is the experimental procedure of the **method of Sweet-Urquhart**?

A. A set of glow curves is recorded using different heating rates.

Q 289. Which is the experimental condition for getting good results in the application of the **variable heating rate method** of analysis?

A. It is essential to have a good thermal contact between the heating element in the TL apparatus and the thermoluminescent sample.

Q 290. If a thermoluminescent material shows an **isothermal decay** law of the form t^{-1}, what kind of kinetics is this supposed to be?

A. A decay law such as t^{-1} can be attributed to a uniform distribution of energies.

Q 291. In which case of a TL glow curve analysis is a **thermal cleaning** necessary?

A. A thermal cleaning procedure is necessary to eliminate peaks which can disturb the peak under study, i.e. shoulder peaks on the ascending part of the studied peak.

Q 292. Are there any methods based on the TL **peak area** for determining the activation energy?

A. Yes. They have been suggested by May and Partridge, Muntoni *et al.*, Maxia *et al.*, and Onnis-Rucci.

Q 293. When is it possible to use the **peak area** methods?

A. The area methods can be used when the TL peak is well isolated and clean.

Q 294. What is the expression for the activation energy determined using the **area** of a first order peak?

A. The activation energy is given by

$$E = kT\left[\ln\frac{s}{\beta} - \ln\frac{1}{\int_T^{T_\infty} I dT'}\right].$$

Q 295. What is the expression for the activation energy calculated using the **area** of a general order peak?

A. The expression is the following

$$E = kT\left[\ln s - \ln\left(\frac{I}{n^b}\right)\right].$$

Q 296. What is the expression of E using the method based on two different **heating rates**?

A. The expression valid for a first order kinetics is the following

$$E = k\frac{T_{M1}T_{M2}}{T_{M1} - T_{M2}} \ln\left(\frac{\beta_1}{\beta_2}\right)\left(\frac{T_{M2}}{T_{M1}}\right)^2 .$$

Q 297. Who proposed the method based on various heating rates?

A. The method was proposed by **Bohm, Porfianovitch and Booth** independently.

Q 298. Can the **various heating rates** methods be applied to a general order kinetics?

A. Yes. Chen and Winer proposed the following expression, valid for a general order of kinetics:

$$\ln\left[I_M^{b-1}\left(\frac{T_M^2}{\beta}\right)^b\right] = \frac{E}{kT_M} + c.$$

Q 299. Who developed the **isothermal decay** method for a first order kinetics?

A. The method was introduced by Garlick and Gibson.

Q 300. What is the expression for determining the **activation energy** from an **isothermal decay** experiment, in the case of a first order kinetics?

A. The expression is the following

$$E = k\frac{T_1 T_2}{T_1 - T_2}\ln\frac{m_1}{m_2}$$

where T_1 and T_2 are two different constant storage temperatures and m_1 and m_2 the slopes of the decay plots.

Q 301. Can the **isothermal decay** be applied when a general order kinetics is involved?

A. Yes. May and Partridge suggested that we should apply the following expression:

$$I^{\frac{1-b}{b}} = a\left[s\exp\left(-\frac{E}{kT}\right)\right]^{\frac{1-b}{b}} + c\left[s\exp\left(-\frac{E}{kT}\right)\right]^{\frac{1-b}{b}} \cdot t.$$

An iterative procedure will give a straight line when the correct value of the kinetics order b is found.

Q 302. Which are the methods based on the **peak shape**?

A. The methods are the following:

- Method based on the high temperature half width
- Method based on the low temperature half width
- Method based on the total half width

The temperature values related to an isolated peak are shown in the following figure:

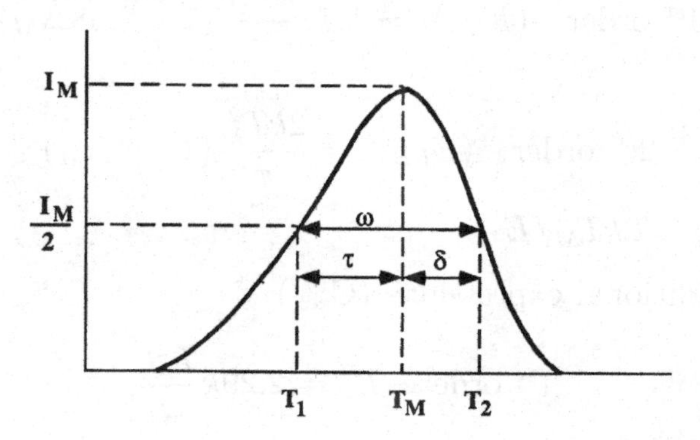

Q 303. What are the expressions for the **peak shape method**?

A. Balarin (B)

$$1^{\text{st}} \text{ order} \quad (E_B)_\omega = \frac{T_M^2}{4998 \cdot \omega}$$

$$2^{\text{nd}} \text{ order} \quad (E_B)_\omega = \frac{T_M^2}{3542 \cdot \omega}$$

Grosswiener (G)

$$1^{\text{st}} \text{ order} \quad (E_G)_\tau = 1.41k\frac{T_1 T_M}{\tau}$$

$$2^{\text{nd}} \text{ order} \quad (E_G)_\tau = 1.68k\frac{T_1 T_M}{\tau}$$

Lushchik (L)

$$1^{\text{st}} \text{ order} \quad (E_L)_\delta = 0.976\frac{kT_M^2}{\delta}$$

$$2^{\text{nd}} \text{ order} \quad (E_L)_\delta = 1.706\frac{kT_M^2}{\delta}$$

Halperin & Braner (E_{HB})

$$1^{\text{st}} \text{ order} \quad (E_{HB})_\tau = 1.72\frac{kT_M^2}{\tau}(1 - 2.58\Delta_M)$$

$$2^{\text{nd}} \text{ order} \quad (E_{HB})_\tau = \frac{2kT_M^2}{\tau}(1 - 3\Delta_M)$$

where $\Delta_M = 2kT_M/E$

Chen's additional expressions (C_{aex})

$$1^{\text{st}} \text{ order} \quad E_\omega = 2.29k\frac{T_M^2}{\omega}$$

$$2^{\text{nd}} \text{ order} \quad E_\omega = 2kT_M\left(1.756\frac{T_M}{\omega} - 1\right)$$

Q 304. Who generalized the **peak shape** methods?

A. R. Chen, in the years 1969–1970, summarized all pre-existing peak shape methods and he also gave expressions for intermediate kinetics order.

Q 305. Which method was used by Chen to obtain the expressions for the **peak shape** procedure?

A. Chen evaluated the coefficients for the first and second order kinetics expressions and then, using the linear interpolation-extrapolation procedure, he also obtained the coefficients for the expressions for the intermediate kinetics orders.

Q 306. On which assumption can be used the **peak shape** methods?

A. The assumption is related to the fact that a glow peak, obtained using a linear and slow heating rate, must be well resolved in the glow curve.

Q 307. What are the **geometrical parameters** of an isolated glow peak?

A. A glow peak is approximated to a triangle and the following parameters are then defined:

T_M is the peak temperature at the maximum,
T_1 and T_2 are respectively the temperatures on either side of T_M corresponding to half intensity,
$\tau = T_M - T_1$ is the half-width at the low temperature side of the peak,
$\delta = T_2 - T_M$ is the half-width towards the fall-off of the glow-peak,
$\omega = T_2 - T_1$ is the total half-width,
$\mu = \delta/\omega$ is the so-called symmetrical geometrical factor.

The following figure shows the different parameters:

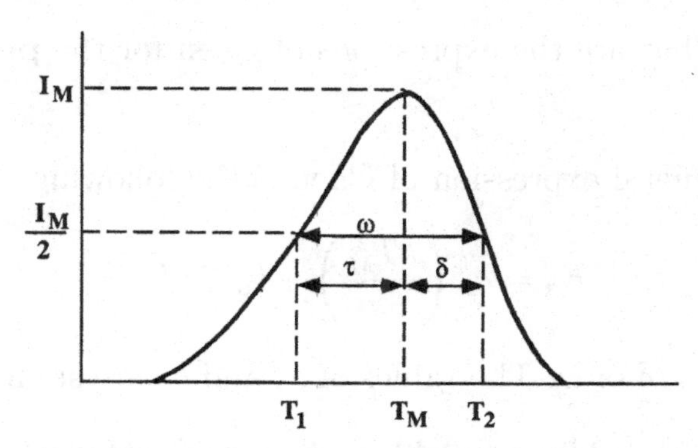

An isolated glow-peak and the temperature parameters.

Q 308. What is the main parameter, related to a peak, used by Chen to obtain the expressions for the **general order kinetics**?

A. Chen obtained the general expressions as a function of the symmetry factor $\mu = \frac{\delta}{\omega}$.

Q 309. What are the values of the **symmetry factor** μ?

A. The symmetry factor ranges between 0.42 and 0.52, where 0.42 is the value for a first order kinetics and 0.52 for a second order.

Q 310. Are intermediate values possible for the **symmetry factor**?

A. Yes. The intermediate values give intermediate kinetics order.

Q 311. What are the expressions of Chen for the **peak shape method**?

A. The general expression of Chen is the following

$$E_\alpha = c_\alpha \left(\frac{kT_M^2}{\alpha} \right) + b_\alpha (2kT_M)$$

where α is τ, δ or ω. The values of c_α and b_α are summarized as:

$$\begin{aligned} c_\tau &= 1.51 + 3.0(\mu_g - 0.42) & b_\tau &= 1.58 + 4.2(\mu_g - 0.42) \\ c_\delta &= 0.976 + 7.3(\mu_g - 0.42) & b_\delta &= 0 \\ c_\omega &= 2.52 + 10.2(\mu_g - 0.42) & b_\omega &= 1 \end{aligned}$$

with

$$\mu_g = 0.42 \quad \text{for first order}$$

$$\mu_g = 0.52 \quad \text{for second order}$$

Q 312. Is it possible to get a plot of the kinetics order as a function of the **symmetry factor**?

A. The plot is given in the following figure:

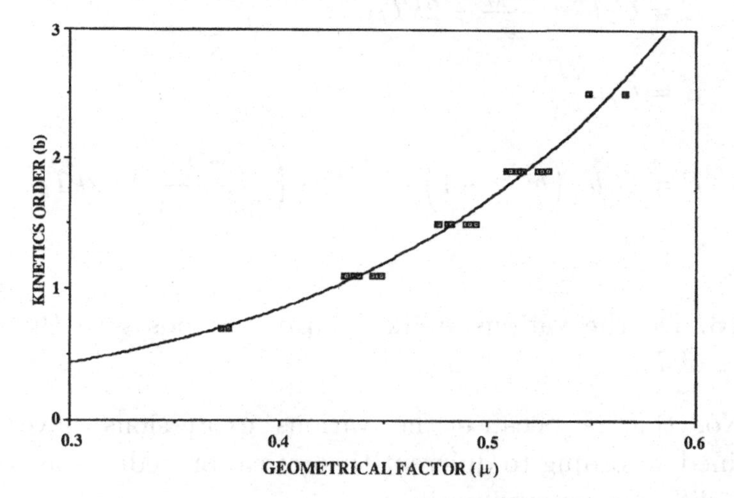

Q 313. Does the Chen's expressions for the **peak shape** have a theoretical basis?

A. No. The Chen's expressions do not have a theoretical basis. Furthermore, because they are a function of the symmetry factor, the kinetics order is absent in the final expressions for the activation energy.

Q 314. Who provided a theoretical foundation to the **peak shape** methods?

A. G. Kitis and V. Pagonis, in 2007, provided a theoretical foundation to the peak shape expressions and gave new expressions for the peak shape method.

Q 315. What are the new **peak shape** expressions obtained by Kitis and Pagonis?

A. The new expressions are

$$E = C_\omega b^{\frac{b}{b-1}} \frac{kT_M^2}{\omega} - 2kT_M$$

$$E = C_\delta b \frac{kT_M^2}{\delta}$$

$$E = C_\tau b \cdot \left(b^{\frac{1}{b-1}} - 1\right) \cdot \frac{kT_M^2}{\tau} - \left(\frac{b^{\frac{1}{b-1}}}{b^{\frac{1}{b-1}} - 1}\right) \cdot 2kT_M$$

Q 316. Do the various **peak shape** methods give the same values of E?

A. No, this is because the various expressions have been obtained according to different theoretical procedures as well as using different approximations.

Q 317. Among the **peak shape** methods, which are more precise?

A. The more precise expressions are the ones of R. Chen.

Q 318. Is there a **reliability criteria** for testing the goodness of the activation energy values obtained by the peak shape methods?

A. A method has been proposed by G. Kitis and it is based on the comparison of the various expressions to the ones of R. Chen.

1st order

$$\left(\frac{E_L}{E_C}\right)_\delta = \frac{0.978K\frac{T_M^2}{\delta}}{0.976K\frac{T_M^2}{\delta}} = 1.002$$

$$\left(\frac{E_G}{E_C}\right)_\tau = \frac{1.41K\frac{T_M T_1}{\tau}}{1.51K\frac{T_M^2}{\tau} - 3.16KT_M} = \frac{T_1}{1.071\left(T_M - 2.09\tau\right)}$$

$$= \frac{T_1}{1.07\left(2.09T_1 - 1.09T_M\right)}$$

$$\left(\frac{E_{HB}}{E_C}\right)_\tau = \frac{\frac{1.72kT_M^2}{\tau}(1 - 2.58\Delta_M)}{\frac{1.51kT_M^2}{\tau} - 3.16kT_M} = 1.139\frac{1 - 2.58\Delta_M}{1 - 2.093\frac{T_M - T_1}{T_M}}$$

Limits:

$$\Delta_M = 0 \rightarrow \frac{1.139}{1 - 2.039\frac{T_M - T_1}{T_M}} = \frac{1.042}{1.915\frac{T_1}{T_M} - 1}$$

$$\Delta_M = 0.1 \rightarrow \frac{0.742}{1 - 2.093\frac{T_M - T_1}{T_M}} = \frac{0.679}{1.915\frac{T_1}{T_M} - 1}$$

$$\left(\frac{0.679}{1.915\frac{T_1}{T_M} - 1}\right)_{\Delta_M = 0.1} \le \left(\frac{E_{\mathrm{HB}}}{E_C}\right)_\tau \le \left(\frac{1.042}{1.915\frac{T_1}{T_M} - 1}\right)_{\Delta_M = 0}$$

$$\left(\frac{E_G}{E_{HB}}\right)_\tau = \frac{1.41k\frac{T_1 T_M}{\tau}}{1.72\frac{kT_M^2}{\tau}\left(1 - 2.58\Delta_M\right)} = \frac{0.8198T_1}{T_M\left(1 - 2.58\Delta_M\right)}$$

Limits:

$$\Delta_M = 0 \rightarrow 0.8198\frac{T_1}{T_M}$$

$$\Delta_M = 0.1 \rightarrow 1.1048\frac{T_1}{T_M}$$

$$\left(0.8198\frac{T_1}{T_M}\right)_{\Delta_M=0} \le \left(\frac{E_G}{E_{HB}}\right)_\tau \le \left(1.1048\frac{T_1}{T_M}\right)_{\Delta_M=0.1}$$

2nd order

$$\left(\frac{E_L}{E_C}\right)_\delta = \frac{1.706\frac{kT_M^2}{\delta}}{1.71\frac{kT_M^2}{\delta}} = 0.998$$

$$\left(\frac{E_G}{E_C}\right)_\tau = \frac{1.68k\frac{T_MT_1}{\tau}}{1.81k\frac{T_M^2}{\tau} - 4kT_M} = \frac{0.7671T_1}{1.8265T_1 - T_M}$$

$$\left(\frac{E_{HB}}{E_C}\right)_\tau = \frac{\frac{2kT_M^2}{\tau}(1-3\Delta_M)}{1.81\frac{kT_M^2}{\tau} - 4kT_M} = 0.917\frac{1-3\Delta_M}{1.83\frac{T_1}{T_M} - 1}$$

Limits:

$$\Delta_M = 0 \to \frac{0.917T_M}{1.83T_1 - T_M}$$

$$\Delta_M = 0.1 \to \frac{0.7T_M}{1.83T_1 - T_M}$$

so that

$$\left(\frac{0.7T_M}{1.83T_1 - T_M}\right)_{\Delta_M=0.1} \le \left(\frac{E_{HB}}{E_C}\right)_\tau \le \left(\frac{0.917T_M}{1.83T_1 - T_M}\right)_{\Delta_M=0}$$

$$\left(\frac{E_G}{E_{HB}}\right)_\tau = \frac{1.68k\frac{T_MT_1}{\tau}}{\frac{2kT_M^2}{\tau}(1-3\Delta_M)} = \frac{0.84T_1}{T_M(1-3\Delta_M)}$$

Limits:

$$\Delta_M = 0 \to 0.84\frac{T_1}{T_M}, \qquad \Delta_M = 0.1 \to 1.2\frac{T_1}{T_M}$$

so that

$$\left(0.84\frac{T_1}{T_M}\right)_{\Delta_M=0} \le \left(\frac{E_G}{E_{HB}}\right)_\tau \le \left(1.2\frac{T_1}{T_M}\right)_{\Delta_M=0.1}$$

Chapter 4

OPTICALLY STIMULATED LUMINESCENCE

Q 319. What is **Optically Stimulated Luminescence (OSL)**?

A. It is a procedure where some irradiated materials, i.e. quartz, is exposed to a steady source of light of appropriate wavelength and intensity able to stimulate luminescence from the material.

Q 320. How do we correlate the luminescence from a material, obtained by using the **OSL** technique, to the dose received by the material?

A. The integral under the curve of the stimulated luminescence signal is a measure of the radiation dose absorbed by the material.

Q 321. Which are the modes of stimulation in **OSL**?

A. They are the Continuous Wave-OSL (CW-OSL), the Linear Modulation-OSL (LM-OSL) and the Pulsed-OSL (POSL).

Q 322. What is the meaning of **CW-OSL** mode?

A. CW-OSL means continuous wave-optically stimulated luminescence.

Q 323. How do we perform the **CW-OSL** mode?

A. It is performed by illuminating an irradiated sample with a constant intensity light source and simultaneously monitoring the luminescence emitted during stimulation.

Q 324. What is the meaning of **LM-OSL**?

A. LM-OSL means linear modulation-optically stimulated luminescence.

Q 325. How do we perform the **LM-OSL** mode?

A. It is performed by increasing the intensity of the stimulation light linearly with time. The procedure is similar to the one used in thermoluminescence, i.e. linear heating rate.

Q 326. What is the meaning of **POSL**?

A. POSL means pulsed-optically stimulated luminescence.

Q 327. How do we perform the **POSL** mode?

A. It is performed using a light stimulation source pulsed at a particular modulation frequency and with a pulse width appropriate to the lifetime of the luminescence.

Q 328. There is the simplest model for **OSL**?

A. The simplest model for OSL involves one electron trap, one hole trap acting as a recombination center and charge transport through the conduction band.

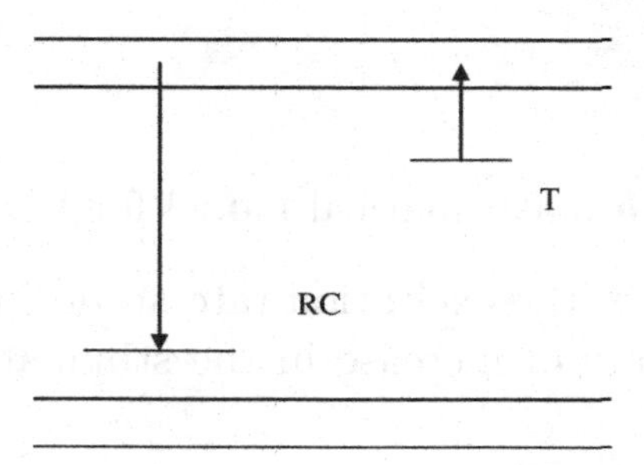

Q 329. What are the rate equations describing the charge flow in the simplest **OSL** model?

A. The rate equations are

$$\frac{dn_C}{dt} = -\frac{dn}{dt} + \frac{dm}{dt}$$

$$I_{CW-OSL} = -\frac{dm}{dt} = -\frac{dn}{dt} = n \cdot f$$

where

n_C = charges in the CB
n = trapped electrons
m = trapped holes
f = wavelength-dependent optical excitation rate = $\Phi\sigma$ where Φ is the illumination flux at wavelength λ and σ is the photoionization cross-section of the defect at wavelength λ.

Assuming the quasi-equilibrium condition and charge neutrality, the final equation is

$$I_{CW-OSL} = I_0 \exp\left(-\frac{t}{\tau}\right)$$

where I_0 is the initial CW-OSL intensity and $\tau = 1/f$ is the decay constant.

Q 330. Is there a mathematical model for **LM-OSL** mode?

A. Yes. Increasing the excitation rate according to $\Phi(t) = \gamma \cdot t$, where γ is the rate of increase in the stimulation intensity, we have

$$I_{LM-OSL} = -\frac{dn}{dt} = n_0 \cdot \sigma \cdot \gamma \cdot t \cdot \exp\left(-\frac{\sigma \cdot \gamma \cdot t^2}{2}\right)$$

where
n_0 is the initial concentration of trapped electrons at time $t = 0$
σ is the photoionization cross-section.

Q 331. What is the equation for the **OSL** intensity in case of general order kinetics?

A. The equation is

$$I_{OSL} = I_0 \frac{t}{T}\left[(b-1)\frac{\alpha I_0 t^2}{2T} + 1\right]^{\frac{b}{1-b}}$$

where

I_0 is the initial OSL intensity at time $t = 0$
$(\alpha I_0)^{-1}$ is the time constant of the luminescence decay
T is the duration of the observation
b is a dimensionless positive value ($\neq 0, 1$) representing the kinetics order.

Q 332. Other than the one trap-one recombination center model, are there more complex models for **OSL**?

A. Yes. One model considers a competing deep trap, a second model considers a competing shallow trap and another is based on the competition of a recombination center.

Q 333. What is the equation for the **OSL intensity** when a competing deep trap is considered?

A. The band model in this case is given in the figure below and the

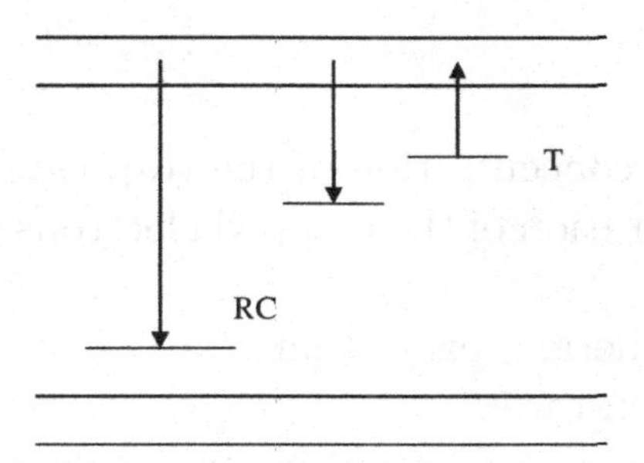

equation is:

$$I_{OSL} = I_{10}\exp\left(-\frac{t}{\tau_1}\right) + I_{20}\exp\left(-\frac{t}{\tau_2}\right)$$

where
I_{10} and I_{20} are the initial luminescence intensity at time $t = 0$
τ_1 and τ_2 are the decay constants.

Q 334. What is the equation for the **OSL intensity** when a competing shallow trap is considered?

A. The band model in this case is given in the figure below and the equation is:

$$I_{OSL} = n_{10} f \exp(-tf) + n_2 p - n_c(N_2 - n_2)A_2$$

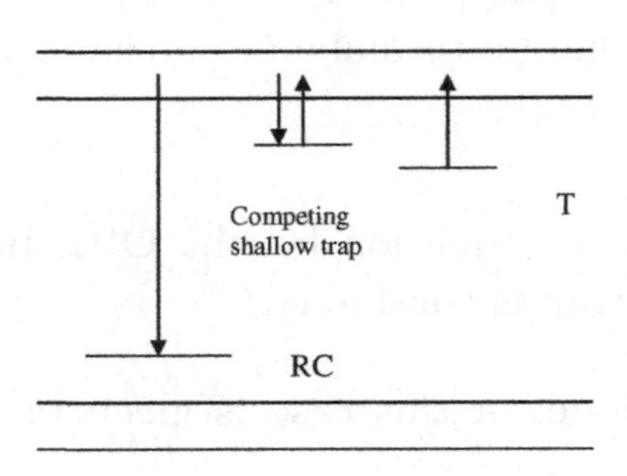

where

n_{10} is the initial concentration of the trapped electrons
n_2 is the concentration of the trapped electrons in the competing shallow trap
p is the rate of thermal excitation
f is the excitation rate
n_c is the electron concentration in the conduction band
N_2 is the concentration of the shallow competing traps
A_2 retrapping probability.

Q 335. What is the equation for the **OSL intensity** when the model of competing recombination center is adopted?

A. The equation is

$$I_{OSL} = I_0 \exp\left(-\frac{t}{\tau}\right).$$

And band model in this case is the following:

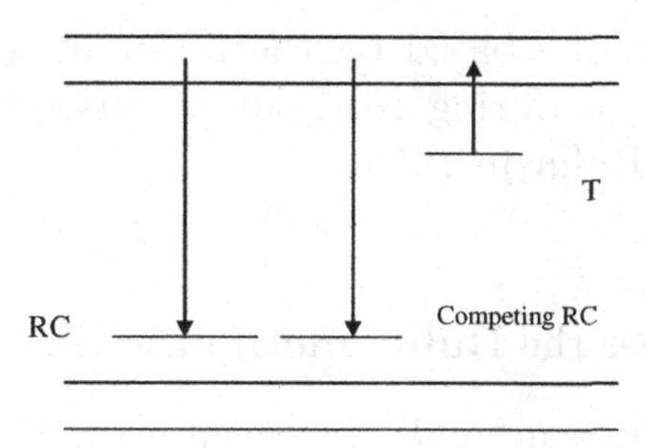

Q 336. What is an important parameter in **OSL measurements**?

A. An important parameter in **OSL** measurements is the wavelength of the excitation light.

Q 337. How is it possible to determine the **wavelength** of the excitation light?

A. The most efficient wavelength can be determined from an excitation spectrum.

Q 338. Is there any relationship between **TL** and **OSL traps**?

A. Only a few possible relationships between TL and OSL traps have been investigated. In $PbWO_4$:Y crystals, OSL defects are different from TL defects. On the contrary, TL and OSL defects seem correlated in $Li_2B_4O_7$:Cu, In phosphor.

Q 339. Can the **OSL** experiments be performed using a linear increase of the stimulating light?

A. Yes. A technique based on the linear increase of the excitation light source during read-out of luminescence has been proposed by E. Bulur in 1966.

Q 340. How does the **Bulur method** work?

A. In the Bulur method the excitation light intensity is increased linearly from zero to a preset value during the read-out of luminescence. In this way the OSL curve can be observed in the form of a peak.

Q 341. What is the model used by Bulur for **OSL** experiments?

A. The band model used by Bulur is shown in the following figure, where it is assumed a single electron trap and a single recombination center.

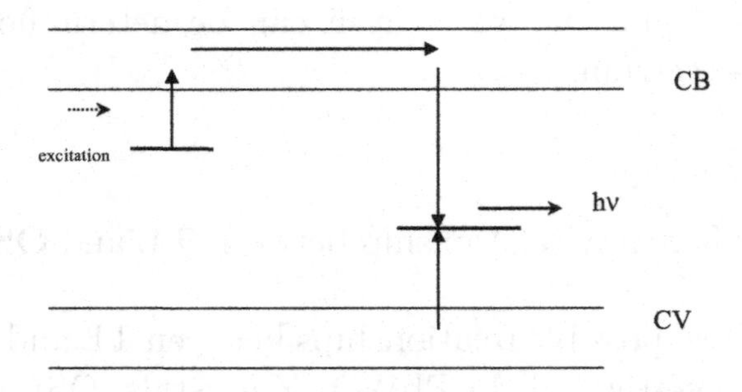

Q 342. What is the first order kinetics equation for the **OSL** process in the Bulur model where the excitation intensity increases linearly with time?

A. The first order equation is the following:

$$\frac{dn}{dt} = -\left[\frac{\alpha I_0}{T}\right] \cdot t \cdot n(t)$$

where $n(t)$ is the number of trapped electrons, αI_0 is the probability of escape of the electrons from the traps at a light intensity I_0 of stimulation light per second, T is the duration of the observation.

Q 343. What is the **luminescence intensity** according to the first order equation given by Bulur, with the excitation intensity increasing linearly with time?

A. The luminescence intensity is given by

$$L(t) = L_0 \frac{t}{T} \exp\left(-\frac{\alpha I_0}{2T} t^2\right)$$

where $L_0 = N_0 \alpha I_0$, and $N_0 = n$ at $t = 0$ is the number of trapped electrons.

Q 344. What is the second order kinetics equation for the **OSL** process in the Bulur model where the excitation intensity increases linearly with time?

A. The second order equation is the following

$$\frac{dn(t)}{dt} = -\frac{\gamma I_0 t n^2(t)}{T}$$

where

$$\gamma I_0 = \frac{\alpha I_0}{N_0}$$

and N_0 is the number of trapped electrons at $t = 0$, αI_0 is the probability of escape of the electrons from the traps at a light intensity I_0 of stimulation light per second, T is the duration of the observation.

Q 345. What is the **luminescence intensity** according to the second order equation given by Bulur, with the excitation intensity increasing linearly with time?

A. The luminescence intensity is given by

$$L(t) = \frac{L_0 \frac{t}{T}}{\left(1 + \frac{\alpha I_0}{2T} t^2\right)^2}$$

where αI_0 is the probability of escape of the electrons from the traps at a light intensity I_0 of stimulation light per second, and T is the duration of the observation.

Q 346. What is the general order kinetics equation for the **OSL** process in the Bulur model where the excitation intensity increases linearly with time?

A. The general order equation is the following

$$\frac{dn}{dt} = -\frac{\gamma I_0}{T} t n^\beta(t)$$

where
$\gamma I_0 = \frac{\alpha I_0}{N_0}$ and N_0 is the number of trapped electrons at $t = 0$, αI_0 is the probability of escape of the electrons from the traps at a light intensity I_0 of stimulation light per second, T is the duration of the observation, and β is a dimensionless positive real parameter $\neq 0, 1$.

Q 347. What is the **luminescence intensity** according to the general order equation given by Bulur, with the excitation intensity increasing linearly with time?

A. The luminescence intensity is given by

$$L(t) = L_0 \frac{t}{T} \left[(\beta - 1) \frac{\alpha I_0 t^2}{2T} + 1 \right]^{\frac{\beta}{1-\beta}} .$$

Chapter 5

LUMINESCENCE DATING

Q 348. What is **luminescence dating**?

A. Luminescence dating is a kind of geochronology based on the measurement of the light emitted by crystalline material contained in archaeological potteries, geological sediments and meteorites in which the radiation energy has been stored during the years.

Q 349. What are the **types of luminescence dating** techniques?

A. A list of the various types of luminescence dating techniques is the following:

- Phototransferred Thermoluminescence (PTTL)
- Thermoluminescence (TL)
- Optically Stimulated Luminescence (OSL)

Q 350. Why are we able to date potteries using **luminescence techniques**?

A. A pottery is a matrix of amorphous fired clay in which is embedded a variety of crystalline minerals, such as quartz and feldspars. Such crystals act as hosts for lattice defects and impurities acting as electron traps when such charges are produced by radiation excitation. Exposure to heat or light stimulation provokes the release of electrons from traps and then the electrons recombine with charges of the opposite sign in the recombination centers. The light emitted during the recombination process is called thermoluminescence (TL) or optically stimulated luminescence (OSL), according to the used technique. The luminescence signal is a direct measure of the number of trapped electrons and it is proportional, through an accurate calibration, to the absorbed dose received by the pottery during the burial period.

Q 351. What is the principle of **luminescence dating**?

A. The principle of luminescence dating is based on the storage information about the radiation energy absorbed in inorganic crystals as quartz and feldspar contained in pottery.

Q 352. How is the **radiation energy** information stored in a crystal?

A. In the band model of the atoms, electron states exist in the forbidden gap, produced by crystal lattice defects and impurities. During the first heating of a ceramic object, all the previously stored information is deleted and so the archaeological watch is set to zero. Over the years the object stores the energy of the absorbed natural radiation. When the object is heated for the TL dating process this energy is released in terms of thermoluminescence light. The emitted thermoluminescence light is then correlated to the absorbed radiation dose and therefore to the archaeological age.

Q 353. What is the origin of the **natural radiation** absorbed by an archaeological object?

A. The natural dose absorbed by a sample is due to alpha, beta and gamma radiation. A few percent is also caused by cosmic radiation.

Q 354. What is the mathematical expression for the **age calculation** using the thermoluminescence technique?

A.

$$Age(yr) = \frac{TLN}{s \cdot \dot{D}}$$

where
TLN is the natural thermoluminescence
s is the sensitivity to the radiation of the thermoluminescent material
$\dot{D}$ is the annual dose rate received by the sample from the surrounding minerals and cosmic radiation

Q 355. How to express the **annual dose rate**?

A. It is given by

$$\dot{D} = \dot{D}_\alpha + \dot{D}_\beta + \dot{D}_\gamma + \dot{D}_c$$

where
$\dot{D}_\alpha$, $\dot{D}_\beta$, $\dot{D}_\gamma$ and $\dot{D}_c$ are the dose rate, natural dose per year, due to alpha, beta, gamma and cosmic radiations respectively.

Q 356. Considering the various radiation components, how to rewrite the **age expression**?

A.

$$Age(yr) = \frac{TLN}{s_\alpha \dot{D}_\alpha + s_\beta \dot{D}_\beta + s_\gamma \dot{D}_\gamma + s_c \dot{D}_c}$$

where

s_α is the sensitivity for alpha radiation
s_β is the sensitivity for beta radiation
s_γ is the sensitivity for gamma radiation
s_c is the sensitivity for cosmic radiation

Q 357. Are the **sensitivities** of the various radiations related in some way?

A. A good assumption is that

$$s_\beta = s_\gamma = s_c = s$$

and

$$s_\alpha = Ks$$

where K ranges from 0.05 to 0.3.

Q 358. According to the assumption for the sensitivities, how should we rewrite the **age expression**?

A.

$$Age(yr) = \frac{TL/s}{K\dot{D}_\alpha + \dot{D}_\beta + \dot{D}_\gamma + \dot{D}_c}.$$

Q 359. How do we measure the **dose rate** for the different types of radiation?

A. The dose rates of the various radiation components are measured by placing thermoluminescent dosimeters or ionization chambers at the sample sites for a known time.

Q 360. How do we calculate the **absorbed dose per year**?

A. The absorbed dose per year is determined by the evaluation of the concentration of natural radiation nuclides in the archaeological object and the measurement of the terrestrial and cosmic radiation at the archaeological site.

Q 361. What are the techniques to be used for the **annual dose rate** calculation?

A. The techniques to be used for the DR determination include neutron activation, atomic absorption, X-ray fluorescence, inductively coupled plasma mass spectrometry, alpha spectrometry, high resolution gamma spectrometry, flame photometry.

Q 362. Which materials can be dated using **luminescence techniques**?

A. The possible materials are: artificial glasses, burned flint and stones, ceramics and burnt clayware, and geological sediments.

Q 363. What are the limits for the archaeological age using **luminescence technique**?

A. The limits are from 10 to 230,000 years.

Q 364. What is the **paleodose**?

A. Paleodose is the term used for the total dose absorbed by a pottery after last firing.

Q 365. What is the **equivalent dose**?

A. It is another way of calling the total absorbed dose.

Q 366. What are the **thermoluminescence dating techniques**?

A. Several thermoluminescence dating techniques have been developed along the years. They are listed below:

- Quartz inclusion technique
- Fine-grain technique
- Subtraction technique
- Zircon inclusion technique
- Pre-dose dating technique
- Phototransferred TL technique

Q 367. How do we **prepare a sample** for dating?

A. A small part of a sample is powdered and then the grain size is choosen according to the technique to be used.

Q 368. What is the **quartz inclusion technique**?

A. A pottery sherd is gently broken by pressing it in the jaws of a vice but avoid crushing it. In this way the grains of quartz of about 100–200 μm size included in the clay matrix are not broken. The thermoluminescent measurements are made on these quartz grains.

Q 369. What are the advantages of the **quartz inclusion technique**?

A. The advantages are:

- quartz is free of radioactive impurities
- it has a very high thermoluminescent response
- the alpha exposure from the burial soil and the clay matrix of the sherd affects the thin outer skin of the quartz grains only a little

Q 370. What is the **fine-grain technique**?

A. A piece of pottery sherd is broken as it is made for the inclusion method or by scrapping with a metal scrap. Grains in the size range of 1–5 μm are separated from the crushed mass by a sedimentation procedure. With this method the alpha, beta and gamma doses can be evaluated.

Q 371. What is the **pre-dose technique**?

A. The quartz grains after some irradiation exhibit a strong peak at 110°C. Pre-dosing and heating allows for the enhancement of the sensitivity of this peak.

Q 372. What is the advantage of the **pre-dose technique**?

A. The advantage of the pre-dose technique is that it is possible for use with young ceramics, i.e. the last thousand years.

Q 373. What is the **subtraction dating technique**?

A. Both techniques of quartz inclusion and fine-grain are combined and only alpha and beta contributions are measured.

Q 374. What is the advantage of the **subtraction technique**?

A. The advantage is that since the gamma component from the environment is not considered, this technique can be used for dating objects from archaeological museums.

Q 375. What is the **zircon inclusion technique**?

A. This method uses highly radioactive, coarse zircon grains. Grain size of 100 mm are separated from the pottery and the thermoluminescence measured.

Q 376. What is the advantage of the **zircon inclusion technique**?

A. Because the zircon grains are large, they only absorb their own alpha dose. This technique is mainly used for authenticity tests.

Q 377. What is the **phototransferred thermoluminescence (PTTL)** method?

A. The sample is exposed to UV rays after annealing. The UV exposure induces transfer of electrons from deep traps to shallow ones. This method has been applied to zircon, apatite and quartz.

Q 378. What is the **plateau test**?

A. The plateau test allows us to determine the temperature region of the glow curve where the thermoluminescent signal is stable, i.e. it is not affect by thermal decay of the ambient ground warmth.

Q 379. Which are the basis of the **plateau test**?

A. A difference exists between the natural glow curve and the thermoluminescent signal induced by an artificial irradiation. Because of thermal detrapping, the thermoluminescence emitted at low temperatures is unstable and then it is absent in the natural glow curve. On the contrary, an artificially dosed sample shows this low temperature thermoluminescence emission. The ratio between the natural and the artificial thermoluminescence signals, as a function of the temperature, shows a plateau at temperature values larger than 300°C. The plateau identifies the temperature region of the natural glow curve in which the thermoluminescence intensity is unaffected by thermal decay.

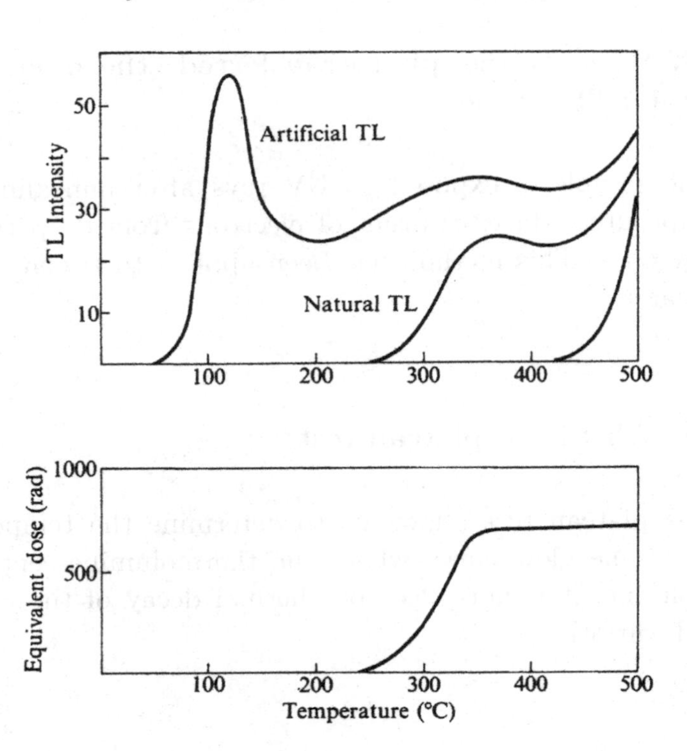

Q 380. What is the **anomalous fading**?

A. It is the phenomenon in which the trapped charges can escape from the traps at a rate much faster than the one expected from the calculated mean lifetime. It is weakly dependent on the temperature.

Chapter 6

MISCELLANEA

Q 381. What is the **radio-luminescence** (**RLL**)?

A. This name indicates a transient emission of light from an irradiated solid upon dissolving it in water or some other solvent. It is a dosimetric method which is not frequently used.

Q 382. What is the meaning of **TSEE**?

A. It means **Thermally Stimulated Electron Emission**.

Q 383. How does **TSEE** work?

A. After having irradiated a sample at low temperature, it is polarized by an electric field applied to the electrodes. During heating, the electron emission is recorded.

Q 384. Does the **TSEE** effect concern the whole sample?

A. The TSEE effect only concerns the surface of the sample.

Q 385. What is the meaning of **TS Cond**?

A. It means **Thermally Stimulated Conductivity**.

Q 386. How does **TS Cond** work?

A. A sample is pressed between two electrodes and then excited at low temperature. The measurements are usually made at constant electric field with linear heating rate and the current is recorded as a function of temperature.

Q 387. What is the meaning of **TSP**?

A. It means **Thermally Stimulated Polarization**.

Q 388. How does **TSP** work?

A. The sample is cooled down under short-circuit conditions and an electric field is applied. If the temperature is raised, then a current is measured versus time and temperature.

Q 389. What is the meaning of **TSC**?

A. It means **Thermo Stimulated Current**.

Q 390. How does **TSC** work?

A. The TSC technique involves the polarization of the sample by a static field at high temperature. The sample is then cooled to a temperature at which the external field is removed and the sample is warmed at a constant rate. A current corresponding to dipole relaxation is recorded as a function of time and temperature.

Q 391. What kind of polarization originated in **TSC** experiment?

A. There are three different types of polarization:

- Orientational polarization
- Space charge polarization
- Interfacial polarization

Q 392. What is the **orientational polarization**?

A. This type of polarization occurs in materials that contain molecular or ionic dipoles.

Q 393. What is the **space charge polarization**?

A. The polarization is due to the presence of excess charge carriers.

Q 394. What is the **interfacial polarization**?

A. The space charge tend to pile up near the interfaces between zones of different conductivities.

Q 395. Who first described a **luminescence phenomenon**?

A. The first description of luminescence was given by Benvenuto Cellini (1565), who observed the luminescence emission from a ruby.

Q 396. Who first gave an accurate description of **luminescence**?

A. A first accurate description of luminescence was given by Robert Boyle in 1663.

Q 397. Who then gave a further description of **luminescence** after the first observation of Boyle?

A. Gian Domenico Cassini observed, in 1676, the luminescence emitted by some materials exposed to the sun light.

Q 398. Who carried out the first experiments on **luminescence**?

A. At the beginning of the XVIII century, Francis Hanksbee carried out some experiments on *phosphorus* in *vacuo*. Also Wall, in the same period, studied the luminescence emission from diamond.

Q 399. Who studied the **luminescence** after Hanksbee and Wall?

A. Further observations and studies on luminescence were carried out by J. H. Pott (1753), J. Canton (1768), J. J. Ferber (1776), J. B. Beccari and B. Wilson (1776), J. J. Le Francais De Lalande (1779), B. Pelletier (1790), L. Donadei (1790),

T. Wedgwood (1792), R. J. Haüy (1801), V. Dessaigues (1809), D. Brewster (1820), A. Becquerel (1839), J. B. Biot (1839), T. J. Pearseall (1839), J. W. Draper (1841), C. Matteucci (1842), E. Becquerel (1843), O. Fiebig (1861), V. Pierre (1866), Kindt (1867), C. Bohn (1867), H. Becquerel (1883) and E. Kester (1899).

Q 400. Who introduced the term **thermoluminescence**?

A. The term thermoluminescence was introduced by E. Wiedemann in 1888.

Q 401. What is the meaning of **ESR**?

A. ESR means *Electron Spin Resonance.*

Q 402. To which fields can **ESR** be applied?

A. ESR can be applied to a large variety of problems in condensed matter physics. In particular, in the areas of retrospective dosimetry and dating applications.

Q 403. What is the time range of the **ESR** dating?

A. ESR dating has a larger time range than any of the dating methods. It is extended from a few thousand to about 2 million years.

Q 404. What are the applications of thermoluminescence in the analysis of **food**?

A. The aim of the thermoluminescence technique applied to foods is to investigate if the foods have been irradiated for sterilization or not.

Q 405. In what way is it possible to check if a certain type of **food** has been irradiated or not?

A. There are two possible techniques in checking the irradiated food: thermoluminescence and optically stimulated luminescence.

Q 406. In what way is thermoluminescence and optically stimulated luminescence different in application in the field of irradiated **food**?

A. To apply thermoluminescence, it is necessary to extract the mineral content of the food; on the contrary, the optically stimulated luminescence does not need this procedure because the sample is not heated during read-out.

Q 407. What are the main characteristics of the glow curves obtained from **irradiated food**?

A. The glow curves from irradiated foods are generally very complex because they are the result of the emitted signals from the various minerals contained in the food. As an example, the following figure shows the glow curves from mint and camomile.

So, in general, the glow curves are a superposition of several peaks, each one corresponding to a certain mineral.

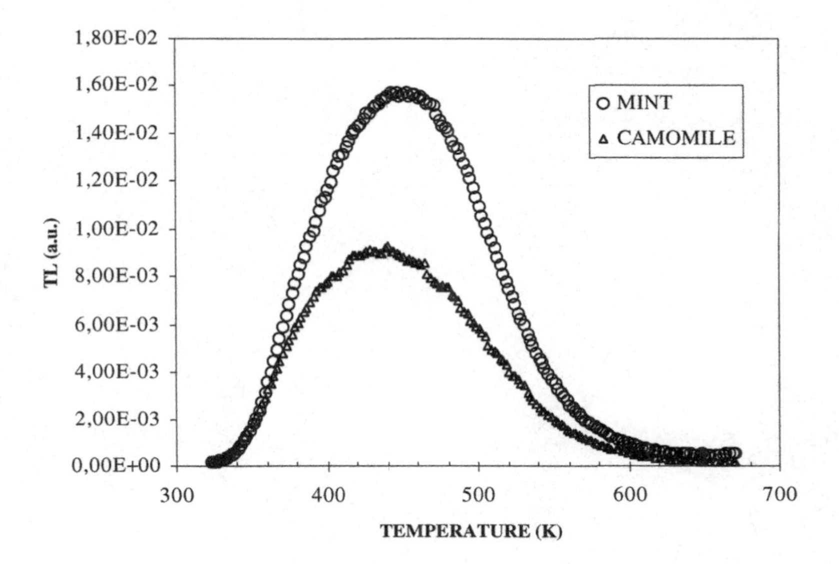

Q 408. How is it possible to carry out kinetics studies on the glow curves produced by **irradiated food**?

A. Because of the complexity of the TL emission, the best way for a kinetics analysis is to apply the McKeever method to get information about the number of possible trapping centers and then to apply the deconvolution method.

Q 409. What kind of deconvolution should be applied to the glow curves from **irradiated food**?

A. In general, the deconvolution should be based on expressions considering a continuous distribution of trapping centers because of the various minerals contained in the foods.

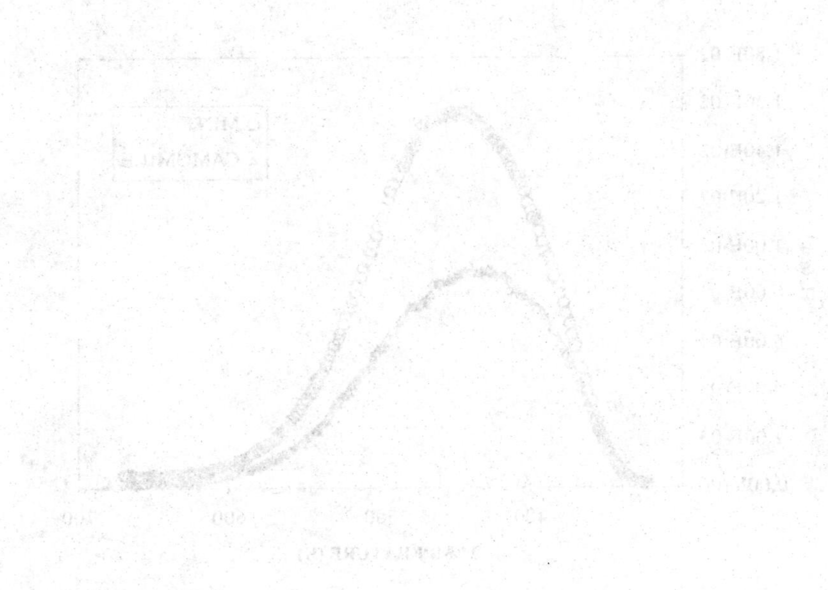

BIBLIOGRAPHY

Adirovitch E. I. A., La formule de Becquerel et la loi elementaire du declin de la luminescence des phosphores crystallin, *J. Phys. Rad.* **17**, 705 (1956).

Aitken M. J., *Thermoluminescence Dating*, Academic Press, London, UK (1985).

Aitken, M. J., *Physics and Archaeology*, Oxford University Press, Oxford, UK (1974).

Alexander C. S., Morris M. F. and McKeever S. W. S., The time and wavelength response of phototransferred thermoluminescence in natural and synthetic quartz, *Radiat. Meas.* **27**, 153 (1997).

Azorin J., Furetta C. and Scacco A., Preparation and properties of thermoluminescent materials, *Phys. Stat. Sol. (a)* **9**, 138 (1993).

Bailey R. M., Smith B. W. and Rhodes E. J., Partial bleaching and the decay form characteristics of quartz OSL, *Radiat. Meas.* **27**, 123 (1997).

Bailiff I. K., Bowman S. G. E., Mobbs S. K. and Aitken M. J., The phototransfer technique and its use in thermoluminescence dating, *J. Electrostat.* **3**, 269 (1977).

Balarin M., Direct evaluation of activation energy from half width of glow peaks and a special monogram, *Phys. Stat. Sol.* **31**, K111 (1975).

Bohm M. and Scharmann A., First order kinetics in thermoluminescence and thermally stimulated conductivity, *Phys. Stat. Sol. (a)* **4**, 99 (1971).

Botter-Jensen L. and Duller G. A. T., A new system for measuring OSL from quartz samples, *Nucl. Tracks Radiat. Meas.* **20**, 549 (1992).

Bowma S. G. E., Phototransferred thermoluminescence in quartz and its potential use in dating, *Eur. PACT J.* **3**, 381 (1979).

Bräunlich P. and Scharmann A., Approximate solutions of Schon's equations for TL and TSC of inorganic photoconducting crystals, *Phys. Stat. Sol.* **18**, 307 (1966).

Bräunlich P., In *Thermally Stimulated Relaxion in Solids*, ed. P. Bräunlich, Spring-Verlag, Berlin (1979).

Bulur E., An alternative technique for optically stimulated luminescence (OSL) experiment, *Radiat. Meas.* **26**(5), 701 (1996).

Burkhardt B. and Piesh E., Reproducibility of TLD systems. A comprehensive analysis of experimental results, *Nucl. Instr. Meth.* **175**, 159 (1980).

Busuoli G., In *Applied Thermoluminescence Dosimetry*, ISPRA Courses, eds. M. Oberhofer and A. Scharmann, Adam Hilger Ltd, Bristol, UK (1981).

Chen I., Theory of thermally stimulated current in hopping systems, *J. Appl. Phys.* **47**(7), 2988 (1976).

Chen R. and Kirsh Y., *Analysis of Thermally Stimulated Processes*, Pergamon Press (1981).

Chen R. and McKeever S. W. S., *Theory of Thermoluminescence and Related Phenomena*, World Scientific (1997).

Chen R. and Winer S. A. A., Effects of various heating rates on glow curves, *J. Appl. Phys.* **41**, 5227 (1970).

Chen R., Glow curves with general order kinetics, *J. Electrochem. Soc.: Solid-State Sci.* **116**(9), 1254 (1969).

Chen R., Methods for kinetic analysis of thermally stimulated processes, *J. Mater. Sci.* **11**, 1521 (1976).

Chen R., On the calculation of activation energies and frequency factors from glow curves, *J. Appl. Phys.* **40**, 570 (1969).

Chen R. and McKeever S. W. S., Characterization of nonlinearities in the dose dependence of Thermoluminescence, *Radiat. Meas.* **23**(4), 667 (1994).

Fillard J. P., Gasiot J. and Manifacier J. C., New approach to thermally stimulated transients: experimental evidence for ZnSe:Al crystals, *Phys. Rev. B* **18**(8), 4497 (1978).

Fleming R. J., Thermally stimulated processes in organic polymers, *Thermochimica Acta* **134**, 15 (1988).

Fleming S. J., *Authenticity in Art*, Institute of Physics, London, UK (1975).

Fleming S. J., *Thermoluminescence Technique in Archaeology*, Clarendon Press, Oxford, UK (1979).

Furetta C. and Gonzalez Martinez P. R., *Termoluminescenza e datazione*, Bagatto Libri Editore, Roma (I) (2007).

Furetta C. and Kitis G., Models in Thermoluminescence, *J. Mater. Sci.* **39**, 2277 (2004).

Furetta C. and Weng P. S., *Operational Thermoluminescence Dosimetry*, World Scientific, Singapore (1998).

Furetta C., *Handbook of Thermoluminescence*, World Scientific, Singapore (2003).

Furetta C., Thermoluminescence, *La Rivista del Nuovo Cimento* **21**(2), (1998).

Garlick G. F. J. and Gibson A. F., The electron trap mechanism of luminescence in sulphide and silicate phosphors, *Proc. Roy. Soc. London* **A60**, 574 (1948).

Grossweiner L. I., A note on the analysis of first order glow curves, *J. Appl. Phys.* **24**, 1306 (1953).

Halperin A. and Braner A. A., Evaluation of thermal activation energies from glow curves, *Phys. Rev.* **117**, 408 (1960).

Horowitz Y. S., *Thermoluminescence and Thermoluminescent Dosimetry*, Vols. I, II and III, CRC Press, USA (1984).

Jonsher A. K., A new approach to thermally stimulated depolarization, *J. Phys. D: Appl. Phys.* **24**, 1633 (1991).

Kitis G. and Pagonis V., Peak shape methods for general order Thermoluminescence glow peaks: a reappraisal, *Nucl. Instr. Meth. B*, in press.

Kitis G., Chen R. and Pagonis V., Thermoluminescence glow peak shape methods based on mixed order kinetics, *Phys. Stat. Sol. (a)*, in press.

Kitis G., Furetta C. and Cruz-Zaragoza E., Reliability criteria for testing the goodness of the activation energy values obtained by the peak shape methods in Thermoluminescence (TL) experiments, *J. Appl. Sci.* **4**(4), 1812 (2005).

Lavergne C. and Lacabanne C., A review of thermo-stimulated current, *IEEE Electrical Insulation Magazine* **9**(2), 5 (1993).

Lewy P. W., Thermoluminescence kinetics in materials exposed to the low dose applicable to dating and dosimetry, *Nucl. Tracks Radiat. Meas.* **10**, 547 (1985).

Lushchik C. B., The investigation of trapping centres in crystals by the method of thermal bleaching, *Sov. Phys. JEPT* **3**, 390 (1956).

Maeta S. and Sakaguchi K., On the determination of trap depth from thermally stimulated current, *Jap. J. Appl. Phys.* **19**(4), 597 (1980).

Marcazzo' J., Santiago M., Spano F., Lester M., Ortega F., Molina P. and Caselli E., Effect of the interaction among traps on the shape of thermoluminescence glow curves, *J. Luminescence* **126**, 245 (2007).

Maxia V., Nonequilibrium thermodynamics of the thermoluminescent process, *Phys. Rev. B* **17**, 3262 (1978).

May C. E. and Partidge J. A., Thermoluminescence kinetics of alpha-irradiated alkali halides, *J. Chem. Phys.* **40**, 1401 (1964).

McKeever S. W. S. and Chen R., Luminescence models, *Radiat. Meas.* **27**(5/6), 625 (1997).

McKeever S. W. S., Larsen N., Botter-Jensen L. and Mejdahl V., OSL sensitivity changes during single aliquot procedures. Computer simulation, *Radiat. Meas.* **27**, 75 (1997).

McKeever S. W. S., Moscovitch M. and Townsend P. D., Thermoluminescence Dosimetry Materials: Properties and Uses, *Nucl. Tech. Publ.*, UK (1995).

McKeever S. W. S., *Thermoluminescence of Solids*, Cambridge University Press (1985).

McKeever S. W. S., On the analysis of complex thermoluminescence glow curves: Resolution into individual peaks, *Phys. Stat. Sol. (a)* **62**, 331 (1980).

McKinley A. F., Thermoluminescence Dosimetry, Adam Hilger Ltd, Bristol, UK (1981).

Muller P., Relationship between thermally stimulated depolarization and conductivity, *Phys. Stat. Sol. (a)* **23**, 165 (1974).

Nambi K. S. V., *Thermoluminescence: Its Understanding and Applications*, Istituto de Energia Atomica, Sao Paolo, Brasil (1977).

Opanowicz A., Comments on the interpretation of thermally stimulated luminescence and conductivity in conductive ZnSe crystals, *Phys. Stat. Sol. (a)* **108**, K47 (1988).

Pagonis V, Kitis G. and Furetta C., *Numerical and Practical Exercises in Thermoluminescence*, Springer (2006).

Randall J. T. and Wilkins M. H. F., Phosphorescence and electron traps I. The study of trap distribution, *Proc. R. Soc. London* **184**, 366 (1945).

Randall J. T. and Wilkins M. H. F., Phosphorescence and electron traps II. The interpretation of long-period phosphorescence, *Proc. R. Soc. London* **184**, 390 (1945).

Rasheedy M. S., On the general order kinetics of the thermoluminescence glow peak, *J. Phys. Condens. Mater.* **5**, 633 (1993).

Scaife B. K. P., On the analysis of thermally stimulated depolarization phenomena, *J. Phys. D: Appl. Phys.* **7**, L171 (1974).

Sunta C. M., Ayta W. E. F., Chubaci J. F. D. and Watanabe S., A critical look at the kinetic models of thermoluminescence: I. First order kinetics, *J. Phys. D: Appl. Phys.* **34**, 2690 (2001).

Sunta C. M., Ayta W. E. F., Kulkarni R. N., Piters T. M. and Watanabe S., Theoretical models of thermoluminescence and their relevance in experimental work, *Rad. Prot. Dos.* **84**, 25 (1999).

Sunta C. M., Ayta W. E. F., Kulkarni R. N., Piters T. M. and Watanabe S., General-order kinetics of thermoluminescence and its physical meaning, *J. Phys. D: Appl. Phys.* **30**, 1234 (1997).

Tyler S. and McKeever S. W. S., Anomalous fading of Thermoluminescence in oligoclase, *Nucl. Tracks Radiat. Meas.* **14**, 149 (1988).

Vallone P., La termoluminescenza dal cinquecento ai giorni nostri, Thesis, Roma, Italy (1991).

Van Dam J. and Marinello G., Methods for *in vivo* dosimetry in external radiotherapy, Garant, Leuven, Belgium (1994).

Vij D. R. (ed.) *Thermoluminescent Materials*, PTR Prentice Hall, New Jersey (1993).

Visocekas R. and Geoffroy A., Tunnelling afterglows and retrapping in calcite, *Phys. Stat. Sol. (a)* **41**, 490 (1977).

Zarand P. and Polgar I., On the relative standard deviation of TLD systems, *Nucl. Instr. Meth.* **222**, 567 (1984).

INDEX